Quantifying Spatial Uncertainty in Natural Resources: Theory and Applications for GIS and Remote Sensing

Edited by

H. Todd Mowrer
and
Russell G. Congalton

CRC Press
Taylor & Francis Group
Boca Raton London New York

CRC Press is an imprint of the
Taylor & Francis Group, an **informa** business

CRC Press
Taylor & Francis Group
6000 Broken Sound Parkway NW, Suite 300
Boca Raton, FL 33487-2742

First issued in paperback 2020

© 2000 by Taylor & Francis Group, LLC
CRC Press is an imprint of Taylor & Francis Group, an Informa business

No claim to original U.S. Government works

ISBN 13: 978-0-367-57901-2 (pbk)
ISBN 13: 978-1-57504-131-5 (hbk)

**Visit the Taylor & Francis Web site at
http://www.taylorandfrancis.com**

**and the CRC Press Web site at
http://www.crcpress.com**

Library of Congress Cataloging-in-Publication Data

Mowrer, H. Todd.
  Quantifying spatial uncertainty in natural resources : theory and applications for GIS and remote sensing / edited by H. Todd Mowrer and Russell G. Congalton.
      p.   cm.
  Based on revised and expanded papers presented at the Second International Symposium on Spatial Accuracy Assessment in Natural Resources and Environmental Sciences.
      ISBN 1-57504-131-6
  1. Geographic information systems—Congresses. 2. Remote sensing—Congresses. 3. Spatial analysis (Statistics)—Congresses. 4. Natural resources—Management—Statistical methods—Congresses. I. Mowrer, H. Todd. II. International Symposium on Spatial Accuracy Assessment in Natural Resources and Environmental Sciences (2nd : 1996 : Fort Collins, Colo.)
  G70.212.C67 1999
  910'.285—dc21   99-14139
                                                                                             CIP

PREFACE

The intended audience for this book consists of both new and experienced practitioners and researchers in spatial uncertainty. We provide practitioners with examples of many applications to explore uncertainty in spatial analyses in remote sensing and geographic information systems. Moreover, this book presents a snapshot of the state of the art of spatial uncertainty assessment, providing theoretical chapters for researchers in this and related disciplines. Many of the 17 chapters are based on revised and expanded papers selected from the 91 presentations at the Second International Symposium on Spatial Accuracy Assessment in Natural Resources and Environmental Sciences. Interestingly, a number of these selected papers also describe continuing research that was presented at the first Spatial Accuracy Symposium, in Williamsburg, Virginia, USA. Additional invited papers have been solicited from participants that reflect work subsequent to the second symposium. All the papers were subjected to rigorous peer review, and those surviving were revised prior to publication. The editors wish to express their deep appreciation to those outside reviewers (and to the authors and coauthors who all reviewed one or more manuscripts) for their professional interest in spatial uncertainty, and for the impressive degree of thought and insight that was apparent in their comments and recommendations:

Outside Reviewers

Barbara P. Buttenfield
David Chojnacky
John B. Collins
Carleton B. Edminster
R.W. Fitzgerald
Rudy M. King
Henry Lachowski
Ronald E. McRoberts
Don Myers

Robin M. Reich
Martin Ritchie
Brian M. Rutherford
Hans Schreuder
James L. Smith
Willem Van Hees
Jill J. Williams
Michael Williams
Curtis Woodcock

The field of spatial uncertainty analysis is rapidly developing into a recognized discipline that combines expertise from geographic information science, remote sensing, spatial and classical statistics, and many others. The final and most important goal of this book is to further this development.

H.T.M. and R.G.C.

THE EDITORS

H. Todd Mowrer is a research scientist with the U.S.D.A. Forest Service, Rocky Mountain Research Station in Fort Collins, Colorado. He received his B.S. and M.S. degrees in Forestry from the University of Illinois, and his Ph.D. in Forest Biometrics from Colorado State University. His entire career has been involved in uncertainty assessment in decision support systems, first in computer simulation models, briefly in artificial intelligence, and for the past 10 years in spatial analysis. His current work focuses on Monte Carlo techniques for assessing uncertainty in geostatistical and computer simulation models in space and time for forest landscape pattern and dynamics. He is Deputy Coordinator of Section 4.01 (Mensuration, Growth and Yield) of the International Union of Forest Research Organizations, and was the Chair of the Second International Symposium on Spatial Analysis in Natural Resources and Environmental Sciences.

Russell G. Congalton has spent much of the last 20 years developing techniques and practical applications for assessing the accuracy of remotely sensed data. This work began in 1979 as an M.S. student in the Department of Forestry at Virginia Polytechnic Institute and State University, continued through his dissertation at the same institution, and has followed him throughout his academic career. Upon graduation, Dr. Congalton was employed as a postdoctorate research associate with the U.S. Army Corps of Engineers Waterways Experiment Station Environmental Lab in 1984. From 1985 to 1991, he held the position of Assistant Professor of Remote Sensing in the Department of Forestry and Resource Management at the University of California, Berkeley. Also during this time he began his relationship with Pacific Meridian Resources which has led to his current role as Chief Scientist. Since 1991, Dr. Congalton has been on the faculty in the Department of Natural Resources at the University of New Hampshire. Currently he is an Associate Professor of Remote Sensing and GIS.

Dr. Congalton has published over 30 peer-reviewed articles and more than 40 conference proceedings papers. He has been a member of the American Society for Photogrammetry and Remote Sensing (ASPRS) since 1979. He was the Conference Director for GIS '87 in San Francisco and was the first National GIS Division Director serving on the National Board of Directors from 1989 to 1991. Currently, he is the Principal Investigator for the Land Cover/Biology Investigation of the GLOBE Program, a project that integrates environmental science with K–12 education sponsored by N.S.F., N.A.S.A., and N.O.A.A.

THE CONTRIBUTORS

Heiko Balzter received his Ph.D. from Justus-Liebig-University in Giessen, Germany, in 1998, where he has been working on spatiotemporal modeling in landscape ecology, particularly with grassland communities. Current research interests focus on the use of radar remote sensing for forest monitoring and for ecological models. He may be contacted at the NERC Institute of Terrestrial Ecology, Section for Earth Observation, Monks Wood, Abbots Ripton, Huntingdon, Cambridgeshire PE17 2LS, UK. Email: Heiko.Balzter@ite.ac.uk

Gary Brand is a Research Forester, USDA Forest Service, North Central Research Station. He has been a member of the Station's Forest Inventory and Analysis Unit since 1995 (previously with the Modeling Unit) and holds an M.F. in Forestry from the University of Minnesota. Gary is interested in the relations between synecological coordinates, ecological classifications, and forest stand dynamics. He can be reached at USDA Forest Service, North Central Research Station, 1992 Folwell Avenue, St. Paul, Minnesota 55108 USA. Email: gbrand/nc@fs.fed.us

Paul Braun is an agricultural scientist, obtaining his doctorate at Giessen University in 1988. He works on agroecology, especially plant population biology and environmental informatics (model validation and spatiotemporal dynamics). He may be contacted at Justus-Liebig-University, Dept. of Biometry and Population Genetics, Ludwigstr. 27, D-35390 Giessen, Germany. Email: paul.braun@agrar.uni-giessen.de

Noel Cressie received the Bachelor of Science degree with First Class Honors in Mathematics from the University of Western Australia and the M.A. and Ph.D. degrees from Princeton University, in 1973 and 1975, respectively. Between 1976 and 1983 he was Lecturer and Senior Lecturer at the Flinders University of South Australia. Since 1983 he has been Professor of Statistics and, after 1993, Distinguished Professor in Liberal Arts and Sciences at Iowa State University, Ames, Iowa. In 1998 he took an appointment as Professor of Statistics at Ohio State University, Columbus, Ohio. His research interests are in statistical image analysis and remote sensing. He may be reached at 404 Cockins Hall, Department of Statistics, The Ohio State University, Columbus, Ohio 43210-1247 USA. Email: ncressie@stat.ohio.ohio-state.edu

D. Richard Cutler is associate professor in the Department of Mathematics and Statistics at Utah State University. He received his Ph.D. in Statistics from University of California, Berkeley in 1989. His research interests include sample survey theory, census adjustment, generalized linear models, and robust estimation. He can be contacted at the Department of Mathematics and Statistics, Utah State University, Logan, Utah 84322-3900, USA. Email: richard@sunfs.math.usu.edu

Ray Czaplewski has worked for 15 years on statistical design and analysis of regional monitoring systems for forests and other environmental features, primarily the U.S.D.A. Forest Service's nationwide Forest Inventory and Monitoring Program. Other accomplishments include statistical design of the 1990 and 2000 global surveys of forests by the Food and Agricultural Organization of the United Nations, and effectiveness monitoring for the President's 1994 Forest Plan for the States of Oregon, Washington, and California. He has specialized in linking sample surveys to remotely sensed data and geographic information systems. He may be contacted at the U.S.D.A. Forest Service, Rocky Mountain Research Station, 240 W. Prospect Rd., Fort Collins, Colorado 80526-2098, USA. Email: rczaplewski/rmrs@fs.fed.us

Thomas C. Edwards, Jr. is a Research Ecologist with the U.S.G.S. Biological Resources Division's Utah Cooperative Fish and Wildlife Research Unit, and is associate professor in the Department of Fisheries and Wildlife, Utah State University. He received his Ph.D. in Wildlife Biology from the University of Florida in 1987. His research interests include wildlife habitat relations modeling and the development of methods for assessing and monitoring biological diversity at large landscape levels. He may be contacted at U.S.G.S. Utah Cooperative Fish and Wildlife Research Unit, Department of Fisheries and Wildlife, Utah State University, Logan, Utah 84322-5210, USA. Email: tce@nr.usu.edu

Francisco Javier Gallego graduated in statistics in Valladolid (Spain), where he became professor in 1986. He moved later to Ispra (Italy) where he works in statistical aspects of remote sensing applications. He can be reached at the Joint Research Centre, I-21020 Ispra (VA) Italy, TP 262. Email: javier.gallego@jrc.it

Michael F. Goodchild is Professor of Geography at the University of California, Santa Barbara and Chair of the Executive Committee, National Center for Geographic Information and Analysis. He received his B.A. at Cambridge University, and his Ph.D. from McMaster University. Among other accomplishments, he has published numerous articles in the field of GIS, six books, and led three symposia on GIS and Environmental Modeling in 1991, 1993, and 1996, and serves on the editorial boards of eight journals and book series. He received the Canadian Association of Geographers award for Scholarly Distinction in 1990 and the Association of American Geographers Award for Outstanding Scholarship in 1996. He may be contacted at NCGIA and Department of Geography, University of California, Santa Barbara, California 93106-4060, USA. Email: good@ncgia.ucsb.edu

Timothy G. Gregoire is currently at the School of Forestry and Environmental Studies, Yale University. Until 1998, he was Associate Professor of Forest Biometrics in the College of Forestry and Wildlife Resources, Virginia Polytechnic Institute and State University, Blacksburg, Virginia, USA. His research concerns both the development of sampling methods for the estimation of forest resources and the modeling of longitudinally and spatially correlated data. He can be reached at Yale University, School of Forestry and Environmental Studies, 360 Prospect Street, New Haven, Connecticut 06511, USA. Email: Timothy.Gregoire@Yale.edu

Andreas Grünig is Project Leader and head of the Advisory Service for Mire Conservation, Swiss Federal Institute of Forest, Snow and Landscape Research. He graduated from the Swiss Federal Institute for Technology, Zurich, and holds a diploma as a scientist (biologist). Andreas is responsible for the Swiss Mire Monitoring Programme. He may be reached at the Swiss Federal Institute for Forest, Snow and Landscape Research, CH-8903 Birmensdorf, Switzerland.

Hubert Hasenauer is Associate Professor at the Institute for Forest Growth Research, Forestry Section, Universität für Bodenkultur in Vienna, Austria. His research interests are forest growth and yield models with special emphasis on individual tree models, biogeochemical models, as well as environmental informatics. He may be reached at Universität für Bodenkultur Wein, Institut für Waldwachstumforschung, Peter Jordan Strasse 82, A-1190, Wein, Austria. Email: hasenau@edv1.boku.ac.at

Rachel Riemann Hershey is a research forester/geographer with the U.S.D.A. Forest Service, Forest Inventory and Analysis Unit, Northeastern Research Station. She received her B.A. in Ecology from Middlebury College, an M.S. in Forestry from the University of New Hampshire, and an M.Phil. in Geography from the London School of Economics. She can be reached at the U.S.D.A. Forest Service, Northeastern Research Station, 11 Camas Boulevard, Newtown Square, Pennsylvania 19073, USA. Email: rriemann/ne_fia@fs.fed.us

Gerard B.M. Heuvelink is a geostatistician with the Faculty of Science, University of Amsterdam. He holds an M.S. in Applied Mathematics from Twente Technical University and a Ph.D. in Geography from Utrecht University. Email: G.B.M.Heuvelink@frw.uva.nl

Chris Hlavka is a research scientist at NASA/Ames Research Center, Moffett Field, California. She received her B.A. in Physics from Pomona College and an M.S. in Statistics from the University of Wisconsin, Madison. She can be contacted at the NASA/Ames Research Center 242-4, Moffett Field, California 94035-1000 USA. Email: chlavka@mail.arc.nasa.gov

Margaret R. Holdaway is a retired Mathematical Statistician who worked for the U.S.D.A. Forest Service, North Central Research Station. She was a member of the Station's Forest Inventory and Analysis Unit until 1999. She is currently employed by the Department of forest Resources, University of Minnesota. She holds an M.S. in Biometry from the University of Minnesota. Her areas of interest are spatial statistics, forest growth modeling, and climatic applications. She can be reached at 1987 Hershel Street, St. Paul, Minnesota, 55113, USA. Email: holda001@tc.umn.edu

Johan A. Huisman is a Ph.D. student at the Faculty of Science, University of Amsterdam. He holds an M.S. in Geography from the University of Amsterdam.

Michael F. Hutchinson is a Senior Fellow at the Australian National University. His primary interest is the spatial and temporal of physical environmental data. He may be reached at the Center for Resource and Environmental Studies, Australian National University, Canberra ACT 0200, Australia. Email: hutch@cres.anu.edu.au

Geoffrey M. Jacquez is President of Bio-Medware, Inc. of Ann Arbor, Michigan, USA. He holds an M.S. in Resource Policy and Law from the University of Michigan, and a Ph.D. in Ecology and Evolution from SUNY Stony Brook, New York. He can be reached at: BioMedware, 516 North State Street, Ann Arbor, Michigan 48104-1236 USA. Email: jacquez@biomedware.com

Michael Köhl is Professor and Chair of Forest Biometrics and Computer Sciences at the Dresden University of Technology, Faculty of Forest, Geo and Hydro Sciences, in Tharandt,

Germany. He graduated from the University of Frieburg, Germany in 1983 and holds a diploma in Forestry and a Ph.D. in biometrics (1986). Michael was lecturer at the Swiss Federal Institute of Technology, Zurich, and was responsible for the statistical and remote sensing methods of the Swiss National Forest Inventory. He is leader of IUFRO S4.11 (Statistics, Mathematics, and Computers). He may be contacted at Dresden University of Technology, Faculty of Forest, Geo and Hydro Sciences, Chair of Forest Biometrics and Computer Science, Wilsdruffer Str. 18, D-01737 Tharandt, Germany. Email: koehl@forst.tu-dresden.de

Wolfgang Köhler is Professor of Biometry and Population Genetics at Justus-Liebig-University and head of the Biometry and Population Genetics Group, Ludwigstr. 27, D-35390 Giessen, Federal Republic of Germany. Email: wolfgang.koehler@agrar.uni-giessen.de

Adam Lewis is the Senior GIS Scientist in the Natural Resource Systems Branch (NRS) of the Department of Conservation and Natural Resources, based in Melbourne, Australia. He has experience in Forest and Land Management, and a Ph.D. from the Australian University. He may be contacted at the Department of Tropical Environment Studies and Geography, James Cook University of North Queensland, Townsville 4811, Australia. Email: adam.lewis@jcu.edu.au

Kim Lowell is a Professor of Forest Geomatics at Laval University in Quebec City, Canada, within the Industrial Chair in Geomatics Applied to Forestry. He received a Ph.D. in Forest Biometrics from Canterbury University, New Zealand in 1985 and was President of the Third International Symposium on Spatial Accuracy in Natural Resources and Environmental Sciences, held at Quebec City, Canada, in 1998. He can be reached at the Faculte de Forestierie et de Geomatique, Centre de Recherche en Geomatique, Local 1353, Pavillon Casault, Cite Universitare, Quebec G1K 7P4, Canada. Email: Kim.Lowell@scg.ulaval.ca

Gretchen G. Moisen is a Research Forester with the Interior West Resource Inventory and Monitoring Unit of the U.S.D.A. Forest Service, Rocky Mountain Research Station. She is currently working toward her Ph.D. in Mathematical Sciences at Utah State University. In addition to her work on uncertainty in spatial databases, Gretchen is involved in comparing the performance of a wide variety of statistical tools for meeting multiple forest inventory objectives. She may be reached at U.S.D.A. Forest Service, Rocky Mountain Research Station, 507 25th Street, Ogden, Utah 84401, USA. Email: gmoisen/rmrs_ogdenfls@fs.fed.us

Robert A. Monserud is Chief Biometrician with the U.S.D.A. Forest Service, with a joint appointment with the Pacific Northwest and Rocky Mountain Research Stations, in Portland, Oregon and Moscow, Idaho, USA, respectively. His current research includes vegetation modeling and paleoclimatic reconstruction for Siberia, forest stand simulation modeling for Austrian forests, modeling genetic and environmental sources of variation for conifers in Idaho, and time-series analysis of stable carbon isotopes in tree rings. He may be reached at Portland Forestry Sciences Lab, 1221 SW Yamhill, Suite 200, Portland, Oregon 97205, USA. Email: monserud/r6pnw_portland@fs.fed.us

Andy Rogowski is currently an Adjunct Professor of Soil Physics at The Pennsylvania State University, and a self-employed consultant in Environmental Science. Born in Poland

and raised in Kenya, he holds an M.S. and Ph.D. degrees in Soil Physics from the Iowa State University. He may be reached at the Department of Agronomy, 116 ASI Building, The Pennsylvania State University, University Park, Pennsylvania 16802-3504, USA. Email: asr2@psu.edu

Hans Jörg Schnellbächer is a research forester with the Inventory Methods Research Group, Swiss Federal Institute of Forest, Snow and Landscape Research. He graduated from University of Freiburg, Germany in 1992 and holds a diploma in Forestry. Hans Jörg provides statistical consulting for the Swiss National Forest Inventory and is responsible for the database management and statistical analysis of the inventory data. He may be reached at Inventory Methods Research Group, Swiss Federal Institute for Forest, Snow and Landscape Research, CH-8903 Birmensdorf, Switzerland.

Lance A. Waller is an Associate Professor of Biostatistics in the School of Public Health at the University of Minnesota. He received his Ph.D. in Operations Research from Cornell University in 1991. He may be contacted at Division of Biostatistics, School of Public Health, University of Minnesota, Minneapolis, Minnesota 55455-0392, USA. Email: lance@muskie.biostat.umn.edu

Michael J.C. Weir is a lecturer at the International Institute for Aerospace Survey and Earth Sciences (ITC), Enschede, The Netherlands. He is mainly responsible for teaching surveying, photogrammetry, and GIS to foresters and other resource specialists from developing countries. He may be contacted at International Institute for Aerospace Survey and Earth Sciences (ITC), P.O. Box 6, Enschede, The Netherlands. Email: weir@itc.nl

Christopher K. Wikle is assistant professor of statistics at the University of Missouri–Columbia. He received B.S. and M.S. degrees in atmospheric science from the University of Kansas in 1986 and 1989, respectively. He obtained an M.S. in statistics at Iowa State University in 1994 and a co-major Ph.D. in both atmospheric science and statistics at Iowa State in 1996. From 1996 to 1998 he was visiting scientist in the geophysical statistics Project at the National Center for Atmospheric Research in Boulder, Colorado. His research interests are in spatio-temporal modeling, hierarchical Bayesian models, and the application of statistics to geophysical processes. He may be contacted at Department of Statistics, 222 Math Science Building, University of Missouri, Columbia, Missouri 65211-4100, USA. Email: wikle@stat.missouri.edu

CONTENTS

INTRODUCTION: THE PAST, PRESENT, AND FUTURE OF SPATIAL UNCERTAINTY ANALYSIS

The Coming-of-Age of Uncertainty Analysis

Increasing awareness and acceptance of spatial uncertainty in natural resource analyses may be attributed primarily to technical and social developments. Technically, the widespread proliferation of geographic information systems (GIS) over the past decade was fueled, in part, by continuing improvements in computer technology, the availability of data from satellite and other remotely sensed sources, and the ability to cheaply georeference locations using global positioning system (GPS) receivers. Socially, increased public interest in the management of natural resources has provided an opportunity to assess analyses from alternative perspectives and viewpoints. GIS provided the ability to analyze the impacts of previous management activities, which had been accomplished in prior years with the best aspatial techniques available, from a new spatial perspective. Increased public involvement in the decision process has encouraged natural resource managers to appraise their management alternatives from all viewpoints, including the accuracy of remotely sensed data and the uncertainty in spatial and temporal analyses.

Prior to the past decade, uncertainty analysis was the province of diverse disciplines such as physics (Taylor, 1982) and nuclear reactor design (Ronen, 1988), and a few zealots in natural resources (Freese, 1960; Alonso, 1967; O'Neill, 1971; Congalton et al., 1983; Gertner and Dzialowy, 1984; Mowrer and Frayer, 1986). With the exception of remote sensing, virtually all other uncertainty analyses were aspatial in nature. In this arena, uncertainty assessment involved verification and validation of computer simulation models. In natural resources, it was often performed almost as an afterthought, if any funds and data were left over after the analysis was complete, and results were reported at the very end of symposia, after all the "important" papers had been presented. The GIS and remote sensing communities, however, recognized uncertainty as an inherent factor in every spatial analysis (Lunetta et al., 1991).

From the outset, the emerging spatial analysis community recognized uncertainty as an inherent problem in spatial analyses. However, it is safe to say that most of the past decade has been consumed by educating users in basic science of GIS and remote sensing, competition between different spatial analysis software, and improvement in software capabilities and user interfaces. While scientific uncertainty is now commonly acknowledged (Lemons, 1995), the theory and application of techniques to quantify uncertainty has lagged far behind the general level of technology in GIS and remote sensing. A notable exception to this overall trend was the 1988 initiative of the National Center for Geographic Information and Analysis (NCGIA), that resulted in a book entitled, *Accuracy of Spatial Databases* (Goodchild and Gopal, 1989). This book enumerated a seven-point research agenda which largely dealt with the effects of the continuously variable scale of digital cartographic representations on numerical precision and accuracy. Not much attention was given to error propagation, statistical precision, or bias, and, as with any good work of seminal research, it admittedly raised more problems and issues than it provided solutions. Nonetheless, this effort provided the immeasurable service of sensitizing the spatial analysis community to the existence of uncertainty, thereby setting the stage for future progress.

In 1994, the first International Symposium on the Spatial Accuracy of Natural Resource Data Bases was organized by Dr. James L. Smith in Williamsburg, Virginia, USA (see Congalton, 1994). This initial gathering brought together some 75 researchers and practitioners interested in the field of spatial uncertainty in natural resource analyses, and was part of a general transition in the field of GIS away from the large general purpose meeting, toward more specialized gatherings. Over 250 participants from more than 20 countries attended the second symposium and over 3,000 copies of the proceedings are in circulation. The third symposium was held in 1998 and continued to build on past success. A fourth symposium is slated for the year 2000 in Amsterdam. Despite progress in many areas, we continue to grapple with a number of the same problems and issues, though hopefully at a higher level of complexity and sophistication.

Uncertainty in No Uncertain Terms

One of the factors that makes spatial analysis such an interesting field of endeavor is that so many other disciplines are involved: geodesy, cartography, database management, computer science, photogrammetry, spatial and classical statistics, remote sensing, and surveying, to name a few. All of these have developed with some degree of independence over the years, and many have their own definitions and approaches for dealing with uncertainty. Under the integrating umbrella of GIS, we are faced not only with estimating the various contributions to uncertainty from each of these sources, but also with integrating their disparate terminology. While uncertainty is rather tersely defined in most dictionaries, in common parlance it may imply anything from the slightest scintilla of doubt (the scientific standard of 95% confidence), all the way to a complete lack of conviction and sheer guesswork. As we define it here, **spatial uncertainty** (in attribute values and in position) includes accuracy, statistical precision, and bias in initial values, and in estimated predictive coefficients in statistically calibrated equations used in the analysis. Most importantly, spatial uncertainty includes the estimation of errors (both in position and in attributes) in the final output that result from the propagation of external (initial value) uncertainty and internal (model) uncertainty. Computer simulation models are often an important component of spatial analyses, and may contribute to uncertainties in attributes and position through the same general processes outlined above. Simulation models are often used to project estimates of particular attributes forward in time (e.g., forest growth modeling), but may be used to make spatial estimates at a point (e.g., moisture infiltration and partitioning through a vertical soil profile). Sources of simulation model uncertainty include not only the uncertainties in initial values and statistical estimation processes previously mentioned, but also the underlying validity of the particular model form and overall structure selected to simulate a given natural process.

Uncertainty also means different things in different disciplines. For example, the concept of **numerical precision** arises from computer science and database management, and refers to the exactness or degree of detail with which an *individual* observation is measured (e.g., "double precision" floating point numbers). **Statistical precision** refers to the dispersion of *repeated* observations about their own mean (Marriott, 1990). Regrettably, in the past, authors have often referred to "precision" without clearly defining whether the precision they are referring to is numerical or statistical. **Accuracy**, on the other hand, is widely accepted as the dispersion of estimates about some "true" value (usually measured at some accepted

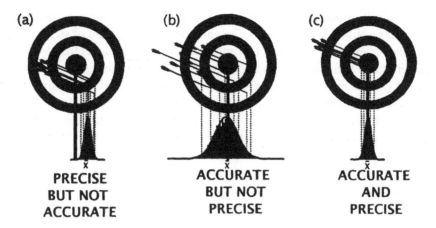

Figure 1. A simplified visualization of statistical precision, accuracy, and bias. In (a), the difference between the central tendency (mean) of the "sample" of arrows, and the center of the bull's-eye represents bias, or systematic error. In many spatial analyses, we are predicting unknown states of nature, and therefore do not know what the "true" value is.

higher order of accuracy). Accuracy includes two components: statistical precision and bias (described below). **Error** may be random or systematic. Random error refers to the difference between an observed value and a predicted or true value (Vogt, 1993). These random deviations are completely the result of chance effects (not human blunders, for example). A common statistic used to describe accuracy is the root mean square error (RMSE), the square root of the quantity, the sum of squares of the errors divided by the number of errors (Slama, 1980). Systematic error, or **bias**, is the consistent difference between the central tendency (mean) of repeated observations, and the known "true" value (see Figure 1). (For an in-depth discussion of these and other definitional concerns, see Mowrer, 1999.)

An Overview of Uncertainty Analysis

The error matrix used to evaluate remote sensing classification (Congalton et al., 1983; Congalton, 1991; Congalton and Green, 1998) is a well-accepted description of uncertainty in primary data. The development of this error matrix is correctly termed accuracy assessment because the classification errors expressed therein are with respect to some reference data (Story and Congalton, 1986). The error matrix is indicative of errors of omission and commission found in the map and is a starting point for various descriptive and analytical statistical techniques. In addition to its use with remotely sensed data, the error matrix or contingency table approach is appropriate to other spatial data applications (e.g., elevation data, soil survey).

Analytical and Empirical Estimates

Two general approaches have been applied in the estimation of uncertainty: analytical (theoretically based) and empirical (observation-based). Each of these have relative advantages and drawbacks, depending on the analysis. In general, analytical approaches are well suited for statistically tractable situations, while empirical uncertainty assessment works for more complex estimation procedures. The analytical development of statistical estimators is very labor-intensive, and is therefore more suited for procedures that do not change. It provides very rapid computation of uncertainty estimates for these static types of analyses. Statistical estimators must be recalculated if the structure of the analysis changes, however. A common empirical approach is Monte Carlo simulation. Key model variables are repeatedly perturbed, the simulation model is run after each perturbation, and multiple, equally probable realizations of the simulation process are created. Uncertainty (e.g., precision) may then be estimated across these realizations. This process relies heavily on computer resources and much less on statistical expertise. Generating an adequate number of Monte Carlo realizations may take hours, days, or even weeks of computer time. While Monte Carlo-based uncertainty assessments place the main burden on the computer, they must also be repeated if model structure changes, and are much more specific in their result. In addition, statistical packages are beginning to incorporate resampling techniques in spatial statistics. These allow some degree of accuracy assessment, though they assume that the model initialization data, used in everyday application, are measured without error. In the "real world" this is seldom the case.

Complications from Correlations

In many of our analyses, we are estimating some unknown state of nature, either spatially or in time. When "true" values (measured at a higher order of accuracy) are not available for comparison, statistical precision may be estimated using the variance about the mean of repeated projections. (Again, this statistical precision should not be confused with numerical precision.) Important components of any estimate of precision are the spatial and interattribute correlations. Multivariate attribute values are described by their mean vector and covariance matrix. Covariance may be calculated from the product of the standard deviations of two variables multiplied by their correlation coefficient. In uncertainty estimation, correlations must be estimated explicitly for analytical estimates, or may be implicitly accounted for through empirical processes; e.g., across Monte Carlo realizations. In computer simulation models, if the same predictive equations are used in multiple sequential projections, correlations must be predicted between attribute values at each projection period. In addition, correlations also accumulate between predicted attributes and estimated predictor coefficients (Mowrer, 1991). Estimation of these correlations has been cited as the source of inaccuracies in analytical procedures (ibid.). In describing uncertainties related to position, correlation is expressed as spatial autocorrelation and cross-correlation. These may be used to form a spatial correlation matrix, as used in cokriging, for example. (For an in-depth discussion of spatial error models, see Heuvelink, 1998.)

Scale

Scale also has important uncertainty implications. When information collected at a larger scale is compared to information represented at a much smaller scale, uncertainty is introduced. Quattrochi and Goodchild (1997) state that, "the effect of generalization is to introduce uncertainty into the representation of a real phenomenon that could only be mapped perfectly at a much larger scale." Digital representations of spatial information are scaleless in that they may be expanded or reduced across a continuous range of ground-to-image ratios. Quattrochi and Goodchild also point out that, rather than stating the metric scale of a digital representation, it is better to define positional accuracy and content: the exactness of the location of the smallest observable object. This brings into play elements of both accuracy (or statistical precision) and numerical precision. The scale of observation also affects the relative degree of accuracy or statistical precision in the observed phenomenon, in that the variance of the means is always smaller than the variance of the individual observations. This implies that coarse grain estimates are likely more precise than equivalent aggregated values resulting from many fine-grain observations (Mowrer, 1989).

Ongoing Problems for Spatial Uncertainty Research

This book addresses a wide variety of problems and issues in spatial uncertainty for natural resources, provides answers to some, and raises new ones for future research. Among those of significance are:

- Uncertainty representation for discrete models of spatial variability; e.g., points, lines, and polygons in vector GIS vs. uncertainty representation for continuous models of spatial variability; e.g., raster GIS. How do we represent uncertainty for models of spatial variability that combine features of both?
- The need to make accuracy assessments more accessible to users of spatial data and model predictions: how can the data collection be more efficient and the results more interpretable?
- The effect of scale (grain and extent) on uncertainty: how aggregation, transformation, or remeasurement of the same or analogous variables across different scales affects uncertainty in final estimates.
- How uncertainty estimates from different disciplines can be combined to provide a composite estimate of uncertainty in final products (e.g., discrete classification of pixels in remotely sensed images vs. continuous variability through spatial interpolation methods, such as kriging).
- How simulation model uncertainty may be integrated with uncertainty in spatial data. How does the spatial variability (spatial autocorrelation and cross-correlation) affect simulation models of temporal phenomena, such as forest growth?
- How can we best develop statistically reliable estimates of uncertainty for procedures that have been accomplished on an *ad hoc* basis in the past. When is it appropriate to use analytical (model-based) processes to estimate uncertainty, and when is it more efficient (statistically or economically) to use empirical (e.g., Monte Carlo, or resampling) methods?
- How can error throughout a project be best evaluated and broken down into its various components (i.e., an error budget) so that each component can be prioritized?

CHAPTER OVERVIEW

Section I. Accuracy Assessment Issues

We have grouped the chapters into two main sections. The first section consists of seven chapters and deals primarily with uncertainty, data quality, sampling, and validation issues in remotely sensed data. Chapter 1 presents an excellent overview of concepts and issues. Chapters 2-7 present a variety of techniques and examples that clearly demonstrate the breadth, complexity, and potential of this critical topic. Positional accuracy is presented in Chapters 2 and 5, while sampling issues are addressed in Chapters 6 and 7. Chapter 3 presents a method for evaluating the factors that affect map error. Readers that are new to this topic may best be served by reading Chapters 1 and 4 first to gain an overview.

In Chapter 1, Michael Goodchild proposes the concept of a data life cycle in which the uses of the data are subjected to various needs and objectives. Therefore, the issues of data quality and accuracy become increasingly important. He presents an introduction to the issues discussed by the other authors in this section and challenges us all to take a broader view of data quality and assessment throughout the life of the data.

In Chapter 2, Adam Lewis and Michael Hutchinson combine data quality with spatial statistics. In this chapter, both positional and thematic accuracy are addressed to define "the fitness of a given data set for a specified use." A simple spatial model is used to represent positional accuracy. Not surprisingly, they conclude that in order to determine the data quality, the use of the data must be specified. Recognizing that it is inefficient to assess the quality of a data set for each and every application, research toward a more generic measure is desirable. Their chapter concludes with an example of modeling a forest pathogen in Southeastern Australia.

Gretchen Moisen, Richard Cutler, and Thomas Edwards present a method of exploring the relationship between accuracy and various map features using a generalized linear mixed model (GLMM) in Chapter 3. In this approach, map features such as topography (e.g., slope and aspect), and vegetation diversity were investigated. The model revealed a strong relationship between accuracy and these map features. There is a great need to continue to explore the factors causing map error in order to improve our maps.

In Chapter 4, Michael Köhl, Hans Schnellbächer, and Andreas Grünig present an example case study in which many of the aspects regarding data quality, sampling, and validation are discussed. The study involves the long-term monitoring of peatlands in Switzerland. Two approaches are presented: the first is a field-sampling approach, and the second is a combined aerial photography and field-sampling approach. The results of the two approaches are compared using a Monte Carlo simulation study. The study showed that the combined approach is superior to the field-sampling approach. Using the aerial photos with the field sampling required only one-fourth of the field samples to obtain the same results.

Chapter 5 by Geoffrey Jacquez and Lance Waller departs from the rest of this section by not using remotely sensed data as part of their application. However, the issue they address, positional uncertainty, is extremely pertinent and therefore the chapter is included here. This chapter contains an excellent overview of the issues surrounding positional accuracy including how this uncertainty affects various statistical measures. This chapter, together with Chapter 2, emphasizes the need to consider positional accuracy when evaluating data quality.

In Chapter 6, Francisco (Javier) Gallego presents a number of examples from various projects of the Commission of the European Community (CEC). These projects cover very large areas in which it is not possible to acquire imagery for the entire area. Instead, double sampling is employed to produce land cover maps, land cover change maps, and valid accuracy assessments. In double sampling, a large sample of units are selected in phase one and some ancillary data are obtained on these samples to determine which samples to use in phase two. Conclusions about the entire area are drawn from these phase two samples.

Finally, Ray Czaplewski continues the discussion of double sampling specifically for accuracy assessment in Chapter 7. This chapter presents a multivariate composite estimator that can be used with a complex sample design in which a large, phase one sample of inexpensive but not necessarily correct data is combined with a smaller, phase two sample of expensive and error-free data. The details on this estimator are presented along with a discussion of its use. The process uses the traditional error matrix with all of its measures of accuracy and analysis techniques.

Section II. Modeling Spatial and Temporal Uncertainty

In the second section, a total of nine chapters address uncertainty in spatial analyses and temporal projections. The first chapter in this section provides a general overview for spatial data management and uncertainty assessment in various agencies throughout the world. The next three chapters investigate geostatistical techniques for modeling spatial data. Chapters 12, 13, and 14 provide insight into discrete modeling of spatial variation, modeling at multiple scales, and reducing uncertainty in temporal modeling. The final three chapters illustrate spatial and temporal modeling for the dynamics of hydrology, vegetation, and precipitation.

In Chapter 8, Michael Weir takes an overall look at how spatial data are acquired by forest management agencies, with particular emphasis on spatial data accuracy as affected by surveying and mapping techniques and the resulting error modeling. Twenty-five agencies were surveyed from throughout the world, though most were in Europe and North America. These agencies were selected based on their use of spatial data in a production mode, rather than research. He concludes that forest management agencies are different from many others involved in similar surveying and mapping work, in that forestry agencies collect their own data and are able to exert a significant influence over data quality. He further recognizes a clear need for agencies to define standards for spatial data acquisition and to use uncertainty assessment of the overall product to define the levels of accuracy and precision that are necessary in those standards.

In Chapter 9, Gerard Heuvelink and Johan Huisman address the first ongoing research problem listed above, that of discrete versus continuous models of spatial variability. In soil science, transitions between soil boundaries have historically been modeled as abrupt. More recently, geostatistics have been applied to include spatial autocorrelation within and across mapping units. Geostatistical techniques, however, lack the ability to simultaneously deal with sharp boundaries and spatial autocorrelation. They recommend the mixed model of spatial variation, because of its ability to deal with either continuous or abrupt changes. Nine simulated "soil maps," with three degrees of discrete versus continuous variability and three increasing levels of spatial autocorrelation, are used to demonstrate the technique.

In Chapter 10, Rachel Riemann Hershey uses the geostatistical techniques of sequential Gaussian conditional simulation and indicator kriging to describe the occurrence of 10 tree species across the state of Pennsylvania, based on individual sample plot locations. Nine of the species exhibit substantial spatial dependence over distances varying from 2 to 100 kilometers. She then investigates the uncertainty in these predictions using sequential Gaussian conditional simulation, which is found to be much more robust to rare species occurrences than the practice of local area averaging. This overall approach allows greater flexibility in the use of species distribution data to capture particular habitat types, for example.

Synecological coordinates are relative indicator values for each species present on a site, determined by its "requirements" for moisture and nutrients (edaphic conditions) and heat and light (climatic conditions). In Chapter 11, Margaret Holdaway and Gary Brand apply ordinary kriging, using both isotropic and anisotropic variograms, as well as universal (trend surface) kriging, to describe the spatial variation for moisture and nutrients across a forested area in Minnesota. Evaluating the procedure using cross-validation and sensitivity analysis, they conclude the anisotropic model works best for nutrients and the trend surface approach for moisture, and that using synecological coordinates reduces the complexity of spatial variation so that broader trends are revealed.

Kim Lowell provides a thought-provoking counterpoint in the next chapter, asking whether discrete polygons or continuous response surfaces are the appropriate way to cartographically model forest stand conditions. Two real-world forestry data sets from Canada are employed to assess this question with regard to stand volume. He investigates between-stand and within-stand interpolability, and compares these to three benchmarks, one of which exhibits a strong, positive spatial autocorrelation, another of which is completely random, and a third synthetic set in which the large values were intentionally clustered. By comparing the behavior of the two real-world stands to that of these three benchmarks, he concludes that in neither real-world case is it appropriate to model the stands as continuous, with the caveat that the polygons may not have completely discrete boundaries.

Christine Hlavka looks at scaling issues in Chapter 13, through an analysis of classified satellite imagery to investigate distortions in the estimated size of small fragments of different classes. Three types of satellite imagery were used, each with different spatial resolution: synthetic aperture radar, Landsat Multi-Spectral Scanner, and Advanced Very High-Resolution Radiometer imagery. She proposes a statistical modeling approach to area estimation, based on models of underlying and observed size distributions of small patches of a highly fragmented cover type.

In Chapter 14, Hubert Hasenauer, Robert Monserud, and Timothy Gregoire suggest the three-stage least squares (3SLS) regression technique as an improved approach to simultaneously estimating biologically related equations in temporal projections of forest growth. This technique accounts for the fact that multiple attributes observed on the same individual are in fact multivariate, instead of independent as traditional modeling techniques assume. This independence assumption is only appropriate when each attribute is observed in an independent sample. Comparing this technique to ordinary least squares (OLS), they found that they were able to simplify the equations by deleting two terms from the OLS fit. By comparing the cross-equation correlations for three predicted variables, they found that the efficiency of 3SLS regression vs. OLS is strongly influenced by the degree of correlation in the full system of equations.

In the following chapter, Andrew Rogowski presents an intriguing case study of soil-water interaction over time and space. His hypothesis is that in the realm of process modeling and GIS, where multiple input parameters are measured at the same point, it is better to model processes at each sample point first, and interpolate outputs later, rather than interpolate first and model later. He demonstrates this for a 1.2 x 1.2 km site in Pennsylvania. Random pedon profile samples were selected at both the site and statewide scale from the soil pedons within the study watershed. The Soil-Plant-Atmosphere-Water (SPAW) water budget simulation model was run on each of the randomly sampled pedon profiles. The resulting estimates of daily output from SPAW were then spatially interpolated using ordinary kriging for variables exhibiting little variation, or sequential indicator simulation to illustrate potential connectivity where spatial patterns exhibited variability. He then follows up with a risk assessment, based on the probabilities calculated over multiple realizations of the sequential indicator simulation, to assess vertical percolation flux, to highlight outliers, to estimate minimum mean absolute deviation, and to select reliable low flux zones. These results are useful to predict sites highly susceptible to groundwater pollution or to custom tailor land use and management practices to water recharge potentials.

In Chapter 16, Paul Braun, Heiko Balzter, and Wolfgang Köhler also look at spatial and temporal dynamics, but for vegetation and plant populations. Using a thistle population recorded as single points, and meadow vegetation measured on a continuum, they apply standard methods of spatial analysis to gain additional insight into the overall dynamics of plant communities, as compared to temporal dynamics alone.

In the concluding chapter, Christopher Wikle and Noel Cressie also look at modeling in space and time through an innovative application of the space-time Kalman filter. The goal of this model is not to predict outside the time frame of the data, but to filter noisy observations and to predict across an entire region at all observation times. This powerfully integrative approach is used to model dynamical contributions from all locations in the spatial domain of interest, to incorporate nonparametric, nonstationary, and anisotropic covariance structures, and to make predictions across unmeasured locations. Using a Kalman filter, they implement a reduced state-space model, using empirical orthogonal functions as the basis set, with the error covariances for the model estimated using the method of moments. The resulting spatially descriptive, temporally dynamic (SDTD) model is then applied to the prediction of precipitation in the South China Sea. Results are estimated for 154 locations over a 10-year period.

CLOSING COMMENT

We hope this introduction will aid the beginner, the analyst, and the scientist in gaining an improved overview of the field of uncertainty assessment in spatial analyses. As we mentioned at the beginning, this book provides techniques for practitioners, as well as theory for researchers. The succeeding chapters address important issues in the list of research problems enumerated above. As this book is intended to reach a variety of levels of expertise and specialization, each reader is encouraged to find his or her own route through these chapters, based on the "road map" we have provided in this introduction. Most importantly, we hope this book will serve to heighten and focus your interest, and will thereby facilitate future innovation in the critically important field of spatial uncertainty assessment.

REFERENCES

Alonso, W. The Quality of Data and the Choice and Design of Predictive Models. Working Paper No. 72, Department of City and Regional Planning and Center for Planning and Development Research, Institute of Urban and Regional Development, University of California, Berkeley, CA, 1967.

Congalton, R.G., R.G. Oderwald, and R.A. Mead. Assessing Landsat classification accuracy using discrete multivariate statistical techniques. *Photogrammetric Eng. Remote Sensing.* 49(12), pp. 1671-1678, 1983.

Congalton, R. A review of assessing the accuracy of classifications of remotely sensed data. *Remote Sensing Environ.* 37, pp. 35-46, 1991.

Congalton, R.G., Ed. Proceedings of the International Symposium on the Spatial Accuracy of Natural Resource Data Bases, May 16-20, 1994, Williamsburg, Virginia, USA. American Society for Photogrammetry and Remote Sensing, Bethesda, MD, 1994.

Congalton, R. and K. Green. *Assessing the Accuracy of Remotely Sensed Data: Principles and Practices.* Lewis Publishers, Boca Raton, FL. (in press), 1998.

Freese, F. Testing accuracy. *For. Sci.,* 6(2), pp. 139-145, 1960.

Gertner, G.Z. and P.J. Dzialowy. Effects of measurement errors on an individual tree-based growth projection system. *Can. J. For. Res.,* 14, pp. 311-316, 1984.

Goodchild, M. and S. Gopal. *Accuracy of Spatial Databases.* Taylor and Francis, London, 1989.

Heuvelink, G.B.M. *Error Propagation in Environmental Modeling with GIS.* Taylor and Francis, London, 1998.

Lemons, J. *Scientific Uncertainty and Environmental Problem Solving.* Blackwell Science, Cambridge, MA, 1995.

Lunetta, R., R. Congalton, L. Fenstermaker, J. Jensen, K. McGwire, and L. Tinney. Remote sensing and geographic information system data integration: Error sources and research issues. *Photogrammetric Eng. Remote Sensing.* 57(6), pp. 677-687, 1991.

Marriott, F.H.C. A Dictionary of Statistical Terms, 5th ed. Longman Scientific and Technical, Essex, England, 1990.

Mowrer, H.T. The Effect of Forest Simulation Model Complexity on Estimate Precision. In *Artificial Intelligence and Growth Models for Forest Management Decisions,* H.E. Burkhart, H.M. Rauscher, and K. Johan, Eds. Publication FWS-1-89, School of Forestry and Wildlife Resources, Virginia Polytechnic Institute and State University, Blacksburg, VA, 1989.

Mowrer, H.T. Estimating components of propagated variance in growth simulation model projections. *Can. J. For. Res.* 21, pp. 379-386, 1991.

Mowrer, H.T. Accuracy (Re)Assurance: Selling Uncertainty Assessment to the Uncertain. In *Spatial Accuracy Assessment: Land Information Uncertainty in Natural Resources,* Proceedings of the Third International Symposium on Spatial Accuracy Assessment in Natural Resources and Environmental Sciences, May 20–22, 1998, Quebec City, Quebec, Canada, Lowell, K. and A. Jaton, Eds. Ann Arbor Press, Chelsea, MI, 1999, pp. 3–10.

Mowrer, H.T. and W.E. Frayer. Variance propagation in growth and yield projections. *Canadian J of Forest Res.,* 16, pp. 1196–1200, 1986.

O'Neill, R.V. Error Analysis of Ecological Models, in *Radionuclides in Ecosystems, Proceedings of the Third National Symposium on Radioecology,* United States Atomic Energy Commission, Oak Ridge, TN, pp. 898-908, 1971.

Quattrochi, D.A. and M.F. Goodchild. *Scale in Remote Sensing and GIS.* Lewis Publishers, Boca Raton, FL, 1997.

Ronen, Y. *Uncertainty Analysis.* CRC Press, Boca Raton, FL, 1988.

Slama, C.C., Ed. *Manual of Photogrammetry,* 4th ed. American Society of Photogrammetry, Falls Church, VA, 1980.

Story, M. and R. Congalton. Accuracy assessment: A user's perspective. *Photogrammetric Eng. Remote Sensing.* 52(3), pp. 397-399, 1986.

Taylor, J.R. *An Introduction to Error Analysis.* University Science Books, Sausalito, CA, 1982.

Vogt, W.P. *Dictionary of Statistics and Methodology.* Sage Publications, Newbury Park, CA, 1993.

Quantifying Spatial Uncertainty in Natural Resources: Theory and Applications for GIS and Remote Sensing

Edited by

H. Todd Mowrer
and
Russell G. Congalton

Section I

Accuracy Assessment Issues

Communicating the Results of Accuracy Assessment: Metadata, Digital Libraries, and Assessing Fitness for Use

Michael F. Goodchild

INTRODUCTION

Accuracy assessment of natural resource data can play a narrowly defined role in the quality control of a range of essential mapping activities, but it can also play a much larger role in supporting the collection, management, storage, use, and archiving of geospatial data. This chapter explores the role of accuracy assessment throughout the *data life cycle,* a term that encompasses all of the stages through which data pass from original observation to eventual filing. There are several reasons for doing this. First, it is clear that any representation of the real world must be an approximation. Hence it is important that the degree of approximation, the amount of information lost when the representation was created, and the errors introduced along the way be maintained as an essential part of the data as they move through the life cycle. They should also be made available to whoever comes into contact with the data, for whatever purpose. Second, there has been a trend recently toward a broadening of interest in the applications of GIS from a fairly narrow stage in the life cycle, where GIS technology is applied to existing databases for the purposes of modeling, analysis, and decision-making, to a concern for the use of GIS throughout the entire life cycle. GIS, it is argued, is not only a desktop technology for the analyst, but also potentially a field technology (Kevany, 1996), and is also important in the later stages of the life cycle when data are shared (Onsrud and Rushton, 1995) and archived (Smith et al., 1996; and see http://alexandria.sdc.ucsb.edu). In this respect there is an increasing degree of linkage and integration between GIS and the related technologies of GPS, remote sensing, wide area networks, and digital libraries.

The chapter is structured as follows. The next sections present an overview of the life cycle, the roles of technologies and stakeholders within it, and the trends that are currently influencing it. The following section discusses metadata, the key element that allows data to travel successfully and usefully through the life cycle and to be shared across a range of applications. The final section looks at the implications of this perspective for accuracy assessment in three contexts: in terms of functions that must be supported; in the need for a hierarchical approach to accuracy assessment; and in the extensions that will be required for existing GIS data models. The concern throughout is with geographic (geospatial) data, defined here as data about specific locations on the surface of the Earth.

THE DATA LIFE CYCLE

Geographic data begin with *direct observation,* either in the field or from above. Measurements are made and transferred to some appropriate medium in analog or digital form (in analog measurement, the results are represented as proportional electrical signals, sound waves, distances on paper, etc.; in digital measurement the representation is coded as a series of digits). Various stages of *interpretation* may occur, as the data are transformed through human interaction. The data may be reformatted, perhaps by conversion from analog to digital form (digitization). They may also be resampled by estimating the values that would have been observed had observations been made at different locations, or aggregated by merging observations. The data may also be converted to *map* form as a general synthesis of knowledge about a defined study area. A *database* is created to allow ready access to the data for the purposes of processing, analysis, and modeling. The data may be *visualized,* modified by various transformations, and used for a range of activities designed to support decisions. In the later stages of the life cycle the data may be stored in some form of *archive,* designed to preserve them for future use, or to allow them to be used for other purposes. *Sharing* of the data may occur between investigators, perhaps across the boundaries of disciplines; many cartographic data, for example, are collected for a wide range of purposes which may or may not be well-defined and well-understood. Eventually the data will become unusable, perhaps because the technology used to store them is no longer supported (punched cards, paper tape, 7-track magnetic tape, 5.25 inch floppy diskettes), or because the medium has deteriorated, or because information necessary for access is no longer available (lost, burned, destroyed).

The data life cycle is sometimes linear, but more often involves feedback loops of varying duration (Figure 1.1). Decisions made on the basis of analysis of data may modify the real world. Transformed data may replace earlier versions. Analysis may lead to further data collection.

The length of the data life cycle is highly variable. Historic data may be available in the form of early maps, and there are major efforts under way to preserve them in digital archives by organizations such as the Library of Congress. On the other hand it is already difficult if not impossible to access early data from remote sensing satellites, such as the early Landsat series, collected as little as 20 years ago. For many GIS projects the data life cycle may be a matter of months, as data are collected and assembled for some specific purpose. But even in these cases the life cycle is likely to be far longer than the duration of the GIS analysis itself. Analysis is increasingly a continuous activity, where GIS technology is applied almost instantaneously as data are collected. This kind of activity is now feasible in areas such as epidemiology, to provide early warning of disease clusters; in climatology; and in agriculture.

In the computer and information sciences, data modeling is defined as the selection of appropriate entities, attributes, and relationships in order to create a useful representation in a digital computer of some complex reality (e.g., Tsichritzis and Lochovsky,

Life cycle stage   **Custodian**

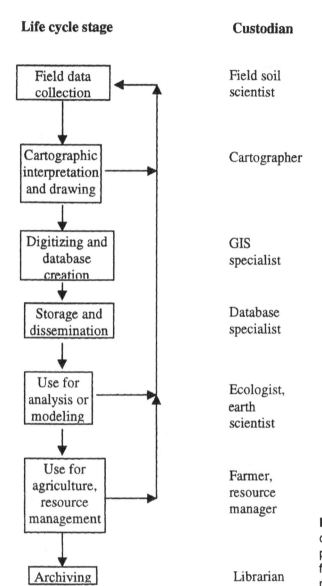

Figure 1.1. The life cycle of a soil
database: an example of the complex patterns of custodianship, transfer, and feedback now common for
many types of spatial data.

1982). Data modeling is particularly important for geographic data because the gap
between the complexity of the real world and the capacity of a digital database is so
great; and because of the large number of options available, particularly for the digital
representation of what is conceived as a geographically continuous surface (Goodchild,
1992). But data modeling is not confined to digital representations—rather, it occurs
at multiple points in the data life cycle as information is reformatted, compressed,
sampled, or otherwise transformed in ways that require a change of basic entities,
attributes, or relationships.

For example, a process that a statistician would call point sampling can also be seen
as a form of data modeling, in which irregularly spaced points capture complex variation that is conceived by the observer as spatially continuous. If the chosen data model

is also to inform its potential users about the information lost when it was created, or the magnitude of the differences between it and the real world, then its entities, attributes, and relationships should include the structures needed to store parameters of error models, variograms, variance surfaces, or whatever is thought to be appropriate in the given instance. These error-related aspects of the data model might attempt to describe both the expected differences between recorded observations and the truth, due to measurement error. They should also describe the uncertainties anticipated when the full continuous surface is interpolated from the point observations. Note that in this example it is immaterial whether the representation is analog or digital—the data modeling issues exist in both cases.

Changes of data model, and decisions about data modeling, can occur at many stages in the data life cycle. They commonly occur:

- during observation;
- during interpretation of observations, such as the interpretation of vegetation boundaries on air photographs which creates linear entities;
- during digitization, such as the replacement of a smooth, analog line on a map with a polyline in a GIS database;
- during resampling associated with projection change or change of spatial resolution; generalization of data; and
- during assembly of results in support of decisions or for archiving.

Rarely does any one individual have a comprehensive view of the entire process of data modeling during the data life cycle. Some data model decisions are constrained by prior decisions made by the designers of observation instruments—the users of remotely sensed data are in this sense locked into a raster data model. Others are driven by communication mechanisms—the user of data expressed on a map, for example, is limited by the data models that can be expressed in map form. It is easy, for example, to create a map representation of a point or a line or a uniform bounded region, but more difficult to express continuous gradations of value using standard cartographic technique. The degree to which the technology of GIS imposes itself on the choice of representations is the subject of much recent discussion and research (Pickles, 1995).

GIS data models have evolved in a field driven largely by the need to represent information that historically has been portrayed in largely analog form on maps. As such, they are ideally suited to data that are static, two-dimensional, and planar, and that express information at a single level of geographic detail. The latter property is particularly problematic during the transition from analog to digital, since metric scale is normally defined from the analog properties of maps, and therefore has no well-defined meaning for digital data (Goodchild and Proctor, 1997). Instead, the stated metric scale of digital data refers to the scale of the map from which the data were digitized; to its thematic contents, since map scale determines the classes of features that can be shown; or to its positional accuracy, since map accuracy standards prescribe expected positional accuracies as a function of scale.

There has been much recent research devoted to extending GIS data models to three dimensions (Raper, 1989; Turner, 1992), time (Langran, 1992), multiple scales

(Buttenfield and McMaster, 1991), and the curved surface of the Earth (Goodchild and Yang, 1992). On the other hand very little work has gone into extensions to accommodate knowledge of accuracy, except as additions to the standard documentation or metadata of a data set (http://fgdc.er.usgs.gov). This is unfortunate, since it means that there is nowhere in a GIS database to store the parameters and other aspects of accuracy that result from the kinds of research discussed in this volume. Moreover, it means that analysis of error propagation; visualization of error; and storage of error properties are difficult to achieve within current software without expensive and poorly linked additions.

Moreover, the frequent modification of data models that occurs along the data life cycle makes it difficult to trace the lineage of a given data item back to the original observations that supported it. For example, a single elevation value in a digital elevation model is likely to have been resampled from points at different densities; to be subject to errors of registration that affect all other points in its neighborhood similarly; and to share other sources of error with its neighbors as well. As a result, errors in geographic data sets tend to be strongly autocorrelated across space, and to have complex structures of spatial dependence. A single value in a geographic database is thus very far from a single, independent observation, and error models for geographic data tend therefore to be relatively complex. In most cases it is impractical to formulate or calibrate models of the various sources of errors that affect geographic databases—instead, error models must often be calibrated by direct comparison between database contents and reality. Much important and potentially useful information about data lineage, such as the locations of registration points, locations of original point observations, and details of interpolations are often lost, either because they are regarded as irrelevant or because of the cartographic tradition of portraying the world as an interpretation, rather than as the result of a set of scientific measurements. In summary, it is unusual to encounter cases in geographic data handling where traditional techniques of measurement error analysis can be applied.

Accuracy in the Life Cycle

Finally, Figure 1.1 shows the frequent changes of custodianship that occur in the data life cycle. Accuracy is defined as the difference between the value of a measurement recorded in the database, and the value of an equivalent measurement from a source of higher accuracy (in the case of geographic data, "higher accuracy" can mean that the reference data are at a more detailed scale, or have been obtained by direct field observation). Unfortunately, a major source of weakness in this definition surrounds the meaning of the term "equivalent." When the custodian of a database assesses accuracy, it is the custodian who is the ultimate judge of whether a reference source is or is not equivalent—whether it is measuring "the same thing." In an extreme case, a custodian who mistook a database of vegetation for a database of geology might assess accuracy against entirely the wrong reference source. Thus it is possible for the accuracy of a database to change, either by improving or by deteriorating, when a database changes hands from one custodian to another. During the data life cycle such changes of custodianship can occur many times. Although the example of geology and vegetation may be extreme, it is relatively common for key elements of information about a data set to be lost in transmission, such as knowledge of the geodetic datum, or the details of a map projection, any of which can lead to inadvertent decrease in accuracy.

In summary, accuracy is a dynamic property of the data life cycle. With reference to Figure 1.1, changes of data model in the left-hand column, and changes of custodian in the right-hand column, produce the potential for significant changes of accuracy as the current representation is compared to the current reference source. Only effective transmission of metadata between custodians, and its effective updating after changes in data model, or repeated reassessment of accuracy, can ensure that adequate information on accuracy is always available to the data's users.

TRENDS AFFECTING THE LIFE CYCLE

This view of the data life cycle and the stages of data modeling is being profoundly affected by changes occurring in society, in computing, and in the application of geographic information technologies. This section examines four of these, and the impacts they are having.

First, GIS is becoming a much more portable technology than previously. In the past, the sheer bulk of the hardware needed for GIS ensured that it would be confined to those stages in the data life cycle that deal with analysis, modeling, and decision-making. GIS is still largely a desk technology, if not a desktop technology, but recently technological developments have allowed it to move much closer to direct support of field observation. GPS is one such development, and the technologies that now support the direct integration of GPS observations with GIS databases. Another is the advent of the laptop computer, and the portable devices exemplified by the Apple Newton—the so-called palmtop computers or personal data assistants. These currently facilitate direct field collection of data, and its uplinking to databases, but will in the future also support more sophisticated analyses such as the direction of point sampling in the field based on statistical principles. They are being implemented in utility companies, forestry agencies, delivery companies, and a host of similar operations that involve acquisition of information on the ground. More efficient dissemination of remotely sensed data will allow them to be integrated with imagery in close to real time for purposes such as fighting wildfire. The set of field GIS applications now includes precision agriculture (Vandenheuvel, 1996), which comes close to locating a fully featured, GPS-linked GIS in the cab of a combine harvester.

Second, the advent of object-oriented thinking has begun to permeate GIS, and to lead to a series of fundamental questions about traditional GIS architectures. One in particular is of importance to accuracy assessment. Most current GIS architectures store location in absolute terms as pairs of coordinates, either on a projected representation of the Earth's surface or in latitude and longitude. While the original observation may be that a given forest boundary occurs approximately 10 m from the edge of a county road, the information actually stored gives measurements of the positions of both boundary and road in absolute terms with respect to the Earth frame, notably the Equator, Poles, and Greenwich Meridian. It has already been argued that many such GIS data items effectively hide their lineage, making error analysis difficult.

Object-oriented thinking favors a distinctly different approach, in which the original measurements are stored and absolute position is regarded as derivative, perhaps computed on the fly when necessary. Such *measurement-based* approaches to GIS are readily supported by object-oriented database designs and programming languages, which allow each location to *inherit* the lineage of its supporting measurements. If these change for some reason, as they often do when there is an opportunity to improve the positional accuracy of some aspect of a database, such as its geodetic control, the dependent data items can be made to update automatically. Similarly, analysis of error is now much easier because the lineage of each data item is not necessarily lost.

Third, the data life cycle as a whole can be seen not as a pipe for the flow of packets of data, but as a communication channel in which the medium of communication is a collection of measurements using some defined data model. In this context it is possible to ask questions regarding the channel's efficiency. Do the chosen data models result in effective communication between the various stakeholders in the data life cycle—do they allow the knowledge gained by the expert soil scientist in the field to be communicated with minimum loss to the eventual user of the data, such as a farmer planting crops? Is it possible to measure the channel's efficiency, and the loss of information resulting from various stages of data modeling? Do the media used at various stages in the cycle result in information loss or other constraints on communication? What needs to be passed through the information channel to make it as easy as possible to realize the benefits of the investment made in the data? How can the channel be broadened so that the data are as useful as possible to the largest number of disciplines?

The trend toward greater use of digital information technologies in the field has already been mentioned. At the other end of the data life cycle, concepts of digital libraries are beginning to influence the archiving of data, and browsing and searching of data resources by researchers and decision-makers. The Alexandria Digital Library project at the University of California, Santa Barbara (Smith et al., 1996; http://alexandria.sdc.ucsb.edu) is one such project directed specifically at the issues associated with building a digital library for geographic data. Its objectives are (1) to develop digital library services for spatially referenced data accessible over the Internet, and (2) to exploit digital technologies in order to bring maps, images, and other forms of spatial data into the mainstream of future libraries.

METADATA AND ACCURACY ASSESSMENT

As data move along the data life cycle they encounter a number of actors who contribute to data's value, make use of data, or process and transform data in some way. The larger the number of actors, and the greater their physical and disciplinary separation, the greater is the challenge in making the data interpretable and useful. Moreover, the ability of the actor to make use of the data depends on the ability to search for suitable data, or more generally to browse.

In the traditional library, the processes of browse and search are supported both by the card catalog (and increasingly by its digital version), and by the order in which books are stacked on shelves. The digital equivalent of the card catalog is *metadata,* or data about data. In addition to the card catalog metaphor, it is useful to think of metadata as containing also the handling instructions for the data, such as details of its format; and also information needed by the user to determine fitness for use. Bretherton (quoted by Smith in Goodchild, 1995) has defined metadata as "information that makes data useful." An important component of metadata is information about the data's quality. Quality is something often taken for granted in the library, and in practice assured by the library's collection-building efforts; in the digital world, more explicit information on quality is often necessary, particularly in the case of geographic data, because of uncertainty about such factors as scale, positional accuracy, date of validity, and lineage.

The digital documentation of geographic data inherent in the concept of metadata is a very active area of research and development in the geographic data community at this time. Two related standards are already in place in the United States as a result of efforts by the Federal Geographic Data Committee (FGDC): the Spatial Data Transfer Standard [SDTS; also known as Federal Information Processing Standard (FIPS) 173; http://fgdc.er.usgs.gov], and the Content Standard for Digital Geospatial Metadata, commonly known as the FGDC standard. Both address issues of importance to the data life cycle, particularly in the area of data quality. SDTS provides extensive detail on the components of data quality, and potential procedures for measurement of appropriate statistics. The general approach aims at ensuring that providers of data have access to appropriate mechanisms for describing what is known about data quality— the so-called "truth in labeling" approach—rather than the establishment of thresholds of quality that must be met.

In attempting to prescribe how data quality should be expressed, however, the details of the standards illustrate a fundamental weakness in the standards enterprise—its inability to evolve in ways that are closely linked to the research frontier. The methods for measuring data quality discussed in the standard reflect the state of research that existed when the standard was written—in the mid-1980s—rather than the kinds of cutting-edge research being discussed in this volume. They are largely silent, for example, on the key issue of spatial structure of error. While it is recommended that accuracy of thematic data sets be described using the error (or confusion) matrix, there is no corresponding recommendation regarding the correlation structure of such errors. So although it is possible to rely on the standard to provide estimates of the uncertainty of classification at a point, it is impossible as a result to estimate the uncertainty associated with any analysis that requires simultaneous examination of more than one point, such as measurement of area, in a digital representation of a multinomial field. Similar problems exist in the suggested methods for describing error in interval/ratio fields, or in discrete point, line, or area objects. Although there is now abundant research on these issues, and although the standard allows for evolution in its specific profiles, there are no comparable mechanisms for modification of the standard itself, which has the effect of freezing progress in the practical implementation of recent data quality research. Of course, the architects of SDTS could hardly have anticipated the wealth of progress that would be made in geographic data quality research in the past decade.

Metadata are often conceptualized as the digital equivalent of a card catalog. But this is inherently limiting, for a number of reasons. First, it limits the digital world to being the analog of the traditional one, rather than encouraging the user to take advantage of the digital world's potential for doing new things, or doing old things in new ways. Second, it makes metadata the exclusive purview of the data producer, who must anticipate possible uses of the data in deciding how best to describe them. In the traditional library this was clearly reasonable, and librarians have developed highly sophisticated ways of cataloging materials in anticipation of the needs of users. But in the digital world there is the opportunity to think of metadata not as something produced by the data's originator or custodian, but as a function of both the producer and the user, or a tool for assessing the degree of fit provided by the data between the producer's assets and the user's needs. In this sense there is no single approach to metadata, but a range of approaches that tries to accommodate to the range of levels of expertise and requirements of the user community. In the case of data quality, for example, metadata might serve one type of user by expressing quality in largely descriptive terms, but might serve another type by providing parameters of a suitable error model, or even an embedded process that when run on the user's system would generate a sample of realizations of the error model. A third type of user might be best served by the creation of a suitable visualization of the effects of error.

From this perspective, it may be better to think of metadata not as the digital equivalent of the card catalog, but as a process of communication between producer and user, in a language understood by both. Since users can be expected to use a variety of languages, the successful producer must match them with a hierarchy of descriptions.

The card catalog metaphor fails in an additional respect which is likely to have profound implications for the geographic data community. In the traditional library, all content occurs within the covers of bound volumes, in convenient, discrete units. The cataloging system is also discrete, since there exists only a finite number of possible subjects, and authors, titles, and subjects can all be ordered along simple dimensions. However, this concept of information granularity fails to survive in the digital world. Data sets can be aggregated into larger units, for example by edgematching tiles to create a seamless view of the world, or by including maps and images in other, larger data sources such as CD encyclopedias. Data can also be usefully disaggregated, when the various digital layers of a simple topographic map are catalogued separately.

The granularity issue becomes more difficult again when discussed in the context of accuracy assessment. In some cases, accuracy may be an attribute of an entire data set, and thus compatible with the card catalog metaphor. In other cases, however, accuracy may be unique to one class of features in a given data set, or may vary by feature, or may be unique to specific subareas. In the latter case, metadata may be better modeled by analogy to the validity map that can be found on many topographic maps—metadata may be a map in its own right, rather than an attribute. In general, it is necessary to think in terms of a hierarchy of possible metadata descriptions of accuracy, ranging from the individual attribute or entity to the seamless database covering the entire Earth.

IMPLICATIONS

Thus far, the chapter has raised a series of issues that arise when accuracy assessment is viewed as a task extending through the entire data life cycle, rather than a one-time process. This final section reviews the implications of this perspective, in three contexts: processes, hierarchical structures, and content.

Processes

From the data life cycle perspective, accuracy assessment is a form of metadata, and an essential part of the communication of information about a data set's fitness for use from one actor in the data life cycle to another, perhaps across substantial barriers of distance and discipline. The accuracy assessment of a given data set must change whenever its data model changes due to some transformation, or whenever other transformations are carried out which invalidate its current accuracy assessment. In some cases such updates can be automated, for example when scale or resolution changes; in other cases updates may require user intervention if there is no simple algorithm available.

Because the uses to which data may be put are likely to vary widely, accuracy assessments must be expressed in terms that are as comprehensive as possible. Not only is the producer or transformer of data under an obligation to describe whatever is known about accuracy that is likely to be of use to eventual users ("truth in labeling")—in addition, potential uses are likely to motivate or even force a more comprehensive assessment than might otherwise have been made. In particular, information on the spatial structure of errors is likely to be useful to a wide range of users, even though it may not be part of the customary apparatus of accuracy assessment.

In principle, the information provided as data move through the data life cycle should be sufficient to allow the user to estimate the uncertainty associated with any product of analysis or modeling. This will include information on the marginal distribution of error in each element of a data set, together with information on the joint distributions of errors at pairs of points. Hunter, Caetano, and Goodchild (1995) and others have described a general and robust strategy for evaluating the effects of error when propagated through GIS operations—the strategy consists of using a suitably calibrated error model to simulate (realize) a sample of possible and equally probable data sets; repeated analysis of each data set; and measurement of the variation across results. This Monte Carlo strategy is sufficiently robust and general to be applied across a full range of GIS operations, and more comprehensive than any analytic approach.

Besides suitable statistics included in metadata, it is desirable that information on data quality be passed between actors using techniques of visualization (Hearnshaw and Unwin, 1994), as the parameters of accuracy assessments and error models may be insufficiently informative to many users. The strategy of Hunter, Caetano, and Goodchild (1995) can be used in this sense also, if the sample of possible data sets is displayed in some appropriate fashion, either by animation or by display in multiple windows (Ehlschlaeger et al., 1997).

Finally, a comprehensive approach to data quality in the data life cycle must include facilities to report the effects of uncertainty in the form of confidence limits, and to update data quality information automatically as the data are passed through transformations and processes (Heuvelink, 1993). Full automation is not likely to be possible, at least in the near future, as the implications of many transformations are not well enough understood. Instead, data quality information may have to be re-worked manually, or simulations may have to be performed to determine the necessary information.

Hierarchy

Although metadata has been seen primarily as the digital analog of the card catalog record, three reasons have been advanced for moving beyond this limited perspective. First, metadata must accommodate the varied experience, vocabulary, and skills of a range of actual and potential users, who range from sophisticated spatial statisticians, armed with geostatistical software, to those with very little familiarity with the deeper issues of geographic information. The statement "digital orthophoto quads have a scale of 1:12,000" may be meaningful to the spatially-aware professional, but may have to be significantly expanded to be meaningful to a grade 10 student. The process of accuracy assessment must lead to a corresponding range of levels of output if it is to be useful to a range of actors along the data life cycle.

Second, structures used to store the results of accuracy assessment in geographic databases must allow for a number of distinct levels of description, depending on the particular stratification of accuracy. It must be possible to describe uncertainty at the levels of attributes, features, feature classes, layers, data sets, and seamless coverages. This suggests that the extension of GIS data models to accommodate information on uncertainty may be far from straightforward. Until it is done, however, the transmission of quality information along the data life cycle will be seriously impeded.

Third, any move to measurement-based GIS requires that the lineage of data be stored explicitly. Instead of latitude/longitude coordinates, the position of a feature in a measurement-based GIS is determined by measurements with respect to some parent feature; absolute coordinates may appear only at the highest level in the hierarchy, which is often associated with geodetic control. Thus data models for measurement-based GIS must also be based in hierarchical structures.

Content

Finally, this chapter has identified various changes that are needed in the content of GIS databases to support an explicit consideration of data quality throughout the data life cycle. First, measurement-based content is clearly more effective than coordinate-based content in allowing us to use standard methods of error analysis, and in dramatically reducing the costs associated with increases in positional accuracy. Second, a comprehensive approach must include not only statistics of the marginal (independent) distributions of data items, but also the joint distributions. Without such information it is impossible to determine the uncertainty associated with any GIS product that requires examination of data items taken more than one at a time—the list of such products includes slope and aspect estimation, determination of polygon area, and a host of other GIS operations, both basic and sophisticated. To support assessment of the accuracy of spatial interpolation, for example, access is needed in GIS data structures to variograms and correlograms, statistics for which there is currently no place in GIS databases.

CONCLUSIONS

This chapter has proposed that accuracy assessment be placed in the broader context of the entire geographic data life cycle. Although there is still much work to be done in refining methods of assessment, and models of error in databases, the value of such work lies ultimately in the degree to which it affects the activities of users who may be very far removed from the originator of the data. Many different actors may be involved in the data life cycle, and many different transformations involving change of data model. At this time, we have only limited understanding of the effects of transformations on accuracy assessments, or the effects of accuracy on the products of GIS operations. On the other hand we do have general strategies which are capable of resolving these issues in robust ways.

The field of GIS is now moving toward such an integrated view of the data life cycle, through the development of search engines for the Web, digital libraries, metadata standards, and related activities. It is important that accuracy assessment be recognized at each stage in the life cycle, and that the results of accuracy assessment be available to all of the actors, in suitable forms.

ACKNOWLEDGMENT

The National Center for Geographic Information and Analysis and the Alexandria Digital Library project are supported by the National Science Foundation, NASA, and the Advanced Research Projects Agency.

REFERENCES

Buttenfield, B.P. and R.B. McMaster, Eds. *Map Generalization: Making Rules for Knowledge Representation,* Longman Scientific and Technical, Harlow, UK, 1991.

Ehlschlaeger, C.R., A.M. Shortridge, and M.F. Goodchild. Visualizing spatial data uncertainty using animation, *Comput. Geosci.,* 23(4), pp. 387–395, 1997.

Goodchild, M.F. Geographical data modeling, *Comput. Geosci.,* 18(4), pp. 401–408, 1992.

Goodchild, M.F. *Report on a Workshop on Metadata held in Santa Barbara, California, November 8, 1995,* http://www.alexandria.ucsb.edu/public-documents/metadata/metadata_ws.html, 1995.

Goodchild, M.F. and J.D. Proctor. Scale in a digital geographic world, *Geograph. Environ. Model.,* 1(1), pp. 5–23, 1997.

Goodchild, M.F. and S. Yang. A hierarchical spatial data structure for global geographic information systems, *Comput. Vision, Graphics, Image Process.: Graphical Models Image Process.,* 54(1), pp. 31–44, 1992.

Hearnshaw, H.M. and D.J. Unwin, Eds. *Visualization in Geographical Information Systems,* Wiley, Chichester, UK, 1994.

Heuvelink, G.B.M. *Error Propagation in Quantitative Spatial Modelling Applications in Geographical Information Systems,* Faculteit Ruimtelijke Wetenschappen, Universiteit Utrecht, 1993.

Hunter, G.J., M. Caetano, and M.F. Goodchild. A methodology for reporting uncertainty in spatial database products, *J. Urban Reg. Inf. Syst. Assoc.,* 7(2), pp. 11–21, 1995.

Kevany, M. The Role of Field Computers in GIS, Paper presented at the 1996 North Carolina Geographic Information Systems Conference, Winston-Salem, February 7–9, 1996.

Langran, G. *Time in Geographic Information Systems*, Taylor and Francis, New York, 1992.

Onsrud, H.J. and G. Rushton, Eds. *Sharing Geographic Information,* Center for Urban Policy Research, New Brunswick, NJ, 1995.

Pickles, J., Ed. *Ground Truth: The Social Implications of Geographic Information Systems,* Guilford Press, New York, 1995.

Raper, J., Ed. *Three-Dimensional Applications in Geographical Information Systems,* Taylor and Francis, New York, 1989.

Smith, T.R. D. Andresen, L. Carver, R. Dolin, and others, A digital library for geographically referenced materials, *Computer*, 29(7), p. 54, 1996.

Tsichritzis, D.C. and F.H. Lochovsky. *Data Models*, Prentice-Hall, Englewood Cliffs, NJ, 1982.

Turner, A.K., Ed. *Three-Dimensional Modeling with Geoscientific Information Systems,* Kluwer Academic Publishers, Dordrecht, 1992.

Vandenheuvel, R.M. The promise of precision agriculture, *J. Soil Water Conserv.,* 51(1), pp. 38–40, 1996.

Chapter 2

From Data Accuracy to Data Quality: Using Spatial Statistics to Predict the Implications of Spatial Error in Point Data

Adam Lewis and Michael F. Hutchinson

INTRODUCTION

We present and demonstrate a simple method for the assessment of data quality, which we loosely define as the fitness of a given data set for a specified use. The focus on data quality, rather than accuracy per se, is regarded as an essential step in the evolution of scientific approaches to the problem of error in spatial databases.

All spatial data sets contain some level of positional error. For data sets in daily use, this error may be in the realm of millimeters (e.g., maps at a scale of 1:1,000, through to kilometers (e.g., maps at the scale of 1:1,000,000); in both cases the data sets are of suitable quality simply because the uses differ; depending on the application, orders of magnitude more or less spatial error may be tolerable.

In applied situations the positional error in data sets, and the tolerance for these levels of error, are often unknown. As a result, most GIS managers are in need of some soundly based indicators of the fitness for use of their key data sets. The extreme cases used above illustrate that a knowledge of the magnitude of error in a spatial data set is, alone, not enough to enable one to determine its fitness for use; the details of the use are all-important.

For point data, spatial error can be modeled as a random process, and several researchers have examined the characteristics of such processes. Most simply a bivariate normal distribution can be used, but other symmetric probability density functions have also been used; e.g., Bolstad et al. (1990). The scaling parameter of the distribution, such as the standard deviation, may be referred to as *epsilon*. In practice, data on the probability distribution of errors are usually lacking; however, it is nonetheless common for mapping agencies and GIS users to quote spatial accuracy in terms of a proportion of points being within a certain distance of the true location. The notion of all or a majority of points falling within a distance, *epsilon*, of the true location, extends in linear features to the *epsilon-band* model of spatial error. These interpretations of spatial error are often quoted in mapping organizations as a standard (Goodchild, 1988; Chrisman, 1997); for instance, "90% of points are within 25 meters of the correct location."

Spatial databases also contain attribute errors; at a given location the value of an attribute recorded in the database may differ from that which would be found in reality. Attribute errors arise because thematic maps, digital elevation models (DEMs), and other spatial data sets are approximations. While polygons map homogeneous units with distinct boundaries, it is well understood that for natural resource data these polygons are heterogeneous. Where continuous spatial variation is represented on a grid or a lattice, as with a DEM, there is a residual "attribute" error in the model, which may be expressed as the root-mean-square (RMS) error.

Both spatial and attribute errors are, typically, spatially autocorrelated; the error in one location is dependent on the errors nearby. This reduces the consequences of error in some circumstances—digital contours do not cross randomly although they may be within each others' *epsilon-band*—but complicates the modeling of errors. The probabilistic point-epsilon model of spatial errors cannot be extended to lines without complication, and models of attribute error require simulation of autocorrelated processes (Goodchild et al., 1992). Kiiveri (1997) addresses this problem by modeling spatial error through a set of randomly chosen basis functions which lead to a "rubber-sheeting" of the data set.

Few studies have addressed the practical implications of errors in spatial data. Ultimately the question is not the absolute error associated with a data set, but whether the data sets available are of adequate *quality* to support a given *use*. Here we develop the relationship between *positional accuracy, use,* and *data quality,* the ultimate aim being to provide a basis from which to make informed statements on fitness-for-use in specific cases.

Initially we develop the theory behind the method and demonstrate its applicability to a data set in East Gippsland, Victoria, Australia. In the latter parts of the chapter we describe further development of the methods and illustrate their application to survey design using a study in Anglesea, Victoria, Australia.

DATA SOURCES AND METHODS

Overview

The analysis is limited to the spatial error of point locations using a simple error model. The impact of the spatial error of the points on the results of overlay with data sets representing slope and elevation is analyzed. We use the R^2 value (the square of the Pearson product-moment coefficient) between the outcome of overlay *given* the spatial error, and the outcome of the overlay *if the points were perfectly placed,* as a measure of the suitability of the point data for overlay with a particular variable; i.e., as a measure of data quality. The study area is a 21,700-hectare catchment in the hilly forested areas of East Gippsland, southeastern Australia (Figure 2.1).

Positional Error of Vegetation Quadrats

We used three sources of location (x,y) data for a set of 155 vegetation quadrats, each measuring 20 by 20 meters. Two of these sources (S_1 and S_2) were suspected to contain some significant positional error, while the third, S_3, can be considered to have negligible positional error relative to the terrain model. All positional errors were estimated relative to

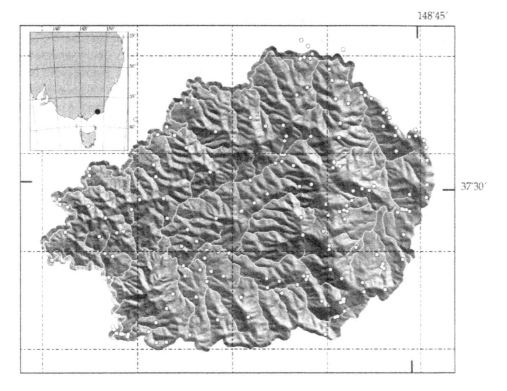

Figure 2.1. East Gippsland study area, vegetation quadrat locations and terrain model (15 meter lattice resolution). The overlaid grid-lines are at 5,000 m intervals.

the terrain model; therefore, positional errors in the terrain model and data-surfaces derived from it are zero.

S_1 quadrat locations were read from a floristic database (Turner and Muek, 1992) in decimal degrees, having been calculated through map projection formulas from map coordinates read from 1:100,000 scale topographic maps. S_1 coordinate error is assumed to be due to incorrect transcription of points onto maps, incorrect reading of coordinates from maps, the low spatial precision of 1:100,000 maps, incorrect transcription of coordinate values, and possibly errors of projection parameters or rounding.

S_2 quadrat locations were digitized directly from the 1:25,000 field maps used in the vegetation survey. These maps show topographic features such as ridge and drainage lines, but not contours. It is reasonable to assume that the quadrat locations are accurately marked onto these maps relative to local topographic features, and that the error in the digitization process is negligible. However, the tortuous lineage of these field maps, which were photographically expanded from 1:63,360 empirical maps compiled several decades earlier to an almost certainly inferior survey base and using unknown projections, raises the specter of significant positional error in the S_2 locations.

S_3 locations were compiled via a separate process which avoids the major sources of positional error inherent in S_1 and S_2, by making use of a more recent and superior topographic base map. The locations of quadrats were marked onto the 1:25,000 standard topographic base map (also used for contour data and therefore terrain modeling), by reference to local topographic features, including relationships to roads, streams, and ridge-lines. The assumption that quadrat locations were originally correctly marked relative to such features is reasonable, since this is how navigation occurs in the field. The remapped locations were then digitized using the coordinate system of the standard topographic base, to give the S_3 locations. S_3 locations (Figure 2.1) are regarded as the true location relative to the terrain model. Errors in the quadrat coordinates of S_1 and S_2 were estimated by comparison with S_3.

Terrain Modeling

A terrain model (Figure 2.1) was interpolated from 1:25,000 contours (10 meter contour interval) for the study area using ANUDEM (Hutchinson 1988, 1989), as implemented in ARCINFO®[1] 7.0 Topogrid. Two attributes of the terrain model, slope and elevation, were estimated at each quadrat location using the ARCINFO® Latticespot command, which applies a bilinear interpolation within the lattice. Slope at each lattice point was estimated as percent rise using the method of Burrough (1986).

Estimation of R^2 Values

The Pearson product-moment correlation coefficient for elevations (or slopes) as read at the true locations (S_3), compared to those read at S_1 and S_2 locations, was calculated.

INITIAL RESULTS

The spatial error associated with the points from S_1 is illustrated in Figure 2.2. S_2 exhibits similar errors with slightly reduced magnitude and no outliers. The *importance* of the positional error, in terms of inducing errors in the estimates of elevation and slope, can be seen from Figure 2.3. For a given source (S_1 or S_2) of positional error the implications of the error are greater for estimates of slope than for elevation. The results given in Figure 2.3 are summarized using R^2 values for slope and elevation between the true locations and the mapped locations, in Table 2.1.

Table 2.1 and Figure 2.3 demonstrate that relatively large errors are introduced into estimates of slope *cf.* elevation, regardless of the source of the quadrat error (S_1 vs. S_2), suggesting that the *type* of variable observed, rather than the specific amount of spatial error in the quadrats, dictates whether the errors introduced will be substantial or negligible.

MODELING THE IMPORTANCE OF SPATIAL ERRORS

The method demonstrated above can only be applied as a post-hoc assessment of the suitability of a given point data set for overlay with a given theme. Furthermore, the true locations of the points must be known, which is unrealistic. Clearly, it is preferable to be able to predict the R^2 values with available data.

---

[1] ARCINFO is a registered trademark of Environmental Systems Research Institute Inc.

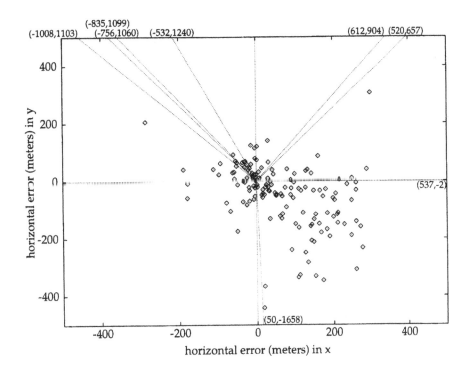

Figure 2.2. Scatter diagram of the horizontal error in x and y observed in quadrat locations from S_1. The diagram indicates that errors in x and y tend to be negatively correlated suggesting an anisotropic error distribution. Errors of more than 500 m in x or y are indicated by the lines radiating from the origin. S_2 data points (not shown) have slightly less error, and do not have extreme values.

In the following we develop the relationships between the expected R^2 value, the positional error of the point data, and the spatial structure of the variable (slope or elevation in this case) used in the overlay. We use the semivariogram (hereafter referred to simply as the variogram) as a measure of spatial structure.

Spatial error is present when $\mathbf{x}_0 \neq \mathbf{x}_1$, $\mathbf{x}_0$ being the true place of observation, and $\mathbf{x}_1$ the location mapped [$\mathbf{x}_0$ and $\mathbf{x}_1$ denote vectors of the form (x,y)]. Let $f(\mathbf{x})$ be a function denoting the probability density of the event $\mathbf{x} = \mathbf{x}_1$ and $Z(\mathbf{x}_1)$ be the actual observation (of slope, insolation, soil depth, or some other spatial variable), while $Z(\mathbf{x}_0)$ is the correct observation (unknown).

We are interested to know, when $\mathbf{x}_0$ is estimated by $\mathbf{x}_1$, what is the expected outcome of observations of some spatial variable, $Z(\mathbf{x}_1)$.

We can state the following:

$$E[Z(\mathbf{x}_1)] = \int Z(\mathbf{x})\, f(\mathbf{x})\, d\mathbf{x} = \int Z(\mathbf{x}_0 + \mathbf{h})\, f(\mathbf{x}_0 + \mathbf{h})\, d\mathbf{h} \qquad (1)$$

where $\mathbf{h}$ is a vector such that $\mathbf{x} = \mathbf{x}_0 + \mathbf{h}$.

It is noteworthy that Equation 1 is a spatial filtering operation undertaken on the surface of Z, with the filter parameters (or kernel) defined by the probability density function of the positional error, $f(\mathbf{x}_0 + \mathbf{h})$. In general, this filtering can be envisaged as a

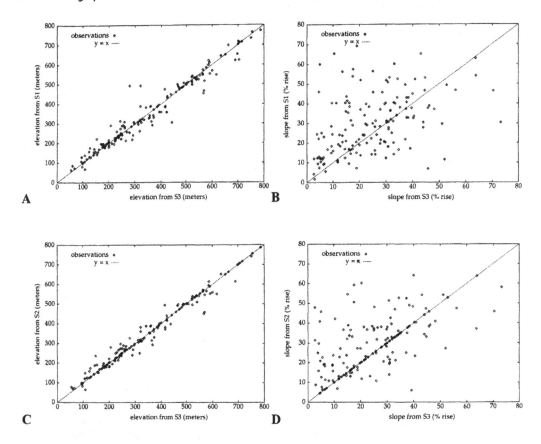

Figure 2.3. Plots of estimates of elevation and slope from quadrat locations from S_1 and S_2, against the correct estimate. Departures from "y=x" indicate errors in the estimate resulting from quadrat location errors: (a) S_1, errors in elevation, (b) S_1, errors in slope, (c) S_2, errors in elevation, (d) S_2, errors in slope. Positional errors in the quadrat locations induce more severe errors in estimates of slope than elevation.

Table 2.1. Correlations Between Observations of Elevation and Slope Taken from S_1 and S_2 Locations, with Observations of the Same Variable, Taken from S_3 Locations.

	Slope (percent rise)	Elevation (meters)
S_1 locations	0.181	0.952
S_2 locations	0.318	0.976

smoothing or blurring of the surface Z, resulting from the spatial error. Equation 1 is also of the form identified by Allen and Starr (1982) as corresponding to observation of $Z(\mathbf{x})$ at a coarser scale, with the scaling function weights being determined by $f(\mathbf{x}_0 + \mathbf{h})$. Accepting this, we would say that spatial error in point locations has similar consequences for the expected value of observation as would viewing at a broader scale. In qualitative terms, the results become blurred. Statistically, the correlation between the observed value $Z(\mathbf{x}_1)$ and the true value $Z(\mathbf{x}_0)$ will drop below 1, unless the spatial error is zero, while the severity of the error will be indicated by the R^2 value between the observed value $Z(\mathbf{x}_1)$, and the true value $Z(\mathbf{x}_0)$.

For a known spatial error $\mathbf{h}$, the covariance, $cov(\mathbf{h})$, between the imperfect observation, taken at location $\mathbf{x}_1 = \mathbf{x}_0 + \mathbf{h}$, and the true observation, taken at $\mathbf{x}_0$, is

$$cov(\mathbf{h}) = E[(Z(\mathbf{x}_0+\mathbf{h}) - m_Z)(Z(\mathbf{x}_0) - \mu_Z)] \tag{2}$$

The correlation coefficient between $Z(\mathbf{x}_1)$ and $Z(\mathbf{x}_0)$ for a known error $\mathbf{h}$ is given by $R(\mathbf{h}) = cov(\mathbf{h}) / \sigma^2$, and

$$R(\mathbf{h})^2 = (cov(\mathbf{h}) / \sigma^2)^2 \tag{3}$$

where μ_Z and σ^2 are, respectively, the mean and variance of the surface Z. $R(\mathbf{h})^2$ may be interpreted as the proportion of the variance in $Z(\mathbf{x}_0)$ accounted for by $Z(\mathbf{x}_1)$.

The covariance is related to the semivariance by $\gamma(\mathbf{h}) = \sigma^2 - cov(\mathbf{h})$, (assuming that Z is a stationary function) so Equation 3 can be restated as:

$$R(\mathbf{h})^2 = (1 - \gamma(\mathbf{h}) / \sigma^2)^2 \tag{4}$$

To progress from the simple case of known error $\mathbf{h}$ to a probability model of error $f(\mathbf{h})$ we calculate the expected value of $R(\mathbf{h})^2$ using simply:

$$E[R^2] = E[R(\mathbf{h})^2] = \int f(\mathbf{h}) \, R(\mathbf{h})^2 \, d\mathbf{h} = \int f(\mathbf{h}) \, (1 - \gamma(\mathbf{h}) / \sigma^2)^2 \, d\mathbf{h} \tag{5}$$

This provides a simple expression to estimate the importance of spatial error in a given situation, requiring only a model of the spatial error of the point observations and a correlogram or variogram of the spatial variable Z. The semivariance, $\gamma(\mathbf{h})$ can be thought of as that part of the total variance which is introduced at lag $\mathbf{h}$.

Equation 5 also illustrates the relationship between the magnitude of the spatial error and the inherent scaling of the variable in determining the importance of the spatial error, discussed earlier. Clearly if $f(\mathbf{h})$ is high only where $\gamma(\mathbf{h})$ is near 0, the error is less important than if $f(\mathbf{h})$ is high over a wide range of $\gamma(\mathbf{h})$ including where $\gamma(\mathbf{h}) \sim \sigma^2$. Forms of $f(\mathbf{h})$ which either have a nonzero mean error, $E[\mathbf{h}] \neq \mathbf{0}$, or which are multimodal, with a peak some distance from $\mathbf{x}_0$ will tend to introduce more significant errors.

Prediction of the Errors in Elevation and Slope Introduced Through Spatial Error in Coordinate Locations

Equation 5 demonstrates that information on the distribution of spatial errors in the point data and a knowledge of the variogram of Z are required to estimate R^2, giving a direct assessment of the *data quality* of the points, for observation of a given variable Z.

If isotropic variograms and error models are assumed vectors can be replaced with appropriate scalars, simplifying Equation 5 to:

$$E[R^2] = \int f(h) \, R(h)^2 \, dh = \int f(h) \, (1 - \gamma(h) / \sigma^2)^2 \, dh \tag{6}$$

where $h = |\mathbf{h}|$.

In the following section this method is applied to the data sets used here. Spreadsheet software was used to manipulate data and to apply mathematical formulas, while variograms were estimated using software developed by the first author. The continuous functions indicated by Equation 6 were approximated using discrete forms with an interval of 40 meters ($h=0,40,...,8000$).

Models of Spatial Error

An isometric error model was adopted. Normal and lognormal distributions were fitted to observed errors using the method of moments (Figure 2.4). The parametric models enable the standard deviation to be used as a parameter to vary the model, allowing sensitivity testing. The assumption of an isometric normal model, $f(h)$, has the further advantage of requiring only one parameter, σ_e. The convention of *epsilon* $= 2\sigma_e$ is adopted, thus in terms of an *epsilon* model of error we can state that 95% of the data points are expected to lie within *epsilon* meters of the mapped point.

A number of authors (for example, Conradsen and Nielson, 1984; Panofsky and Dutton, 1984) have observed that if the errors in x and y are assumed to follow independent identical normal distributions then the preferred model for the magnitude of the errors is a Weibull (Rayleigh) distribution. Preliminary investigations found that, for our data, a normal distribution offered a superior fit.

ASSESSING DATA QUALITY FOR SLOPE AND ELEVATION

The discretized version of Equation 6 was used to model the expected R^2 values for slope and elevation from S_1 and S_2, for values of *epsilon* corresponding to these data sets (*epsilon* $= 384$ m for S_1, and *epsilon* $= 320$ m for S_2).

$$E[R^2] = \sum [F(h+\Delta/2) - F(h\,\Delta/2)][1-\gamma(h)/\sigma^2]^2$$
$$h = \Delta,\ 2\Delta,\ ...,\ 8000$$
$$+ [F(\Delta/2) - F(0)][1-\gamma(0)/\sigma^2]^2 \tag{7}$$

where: $F(h)$ is the cumulative normal probability distribution with standard deviation of *epsilon*/2, truncated to exclude negative values ($F(0) = 0$) and scaled to maintain $F(\infty) = 1$; h is the magnitude of the spatial error, and the lag distance for $\gamma(h)$; Δ is an interval (40 m), σ^2 is the observed variance of Z (slope or elevation, as appropriate). The results, and corresponding observed values, are shown in Table 2.2.

In most practical situations, $\gamma(h)$ is readily estimated for data sets in a GIS, but the precise nature of $F(h)$ will be unknown. To develop some idea of the sensitivity of data quality to $F(h)$, $E[R^2]$ was calculated for a wide range of *epsilon*. Results are shown in Figure 2.5, from which it is clear that the R^2 values for elevation from S_1 and S_2 are closely predicted. R^2 for slope is accurately predicted for S_2, but for S_1 is optimistic, reflecting the extreme positional errors not catered for by the normal distribution (Figure 2.4).

The contrast between Figures 2.5a and 2.5b reinforces our finding that the key factor determining the R^2 between correct and incorrect values of elevation and slope is *not* the spatial error in the observation points, but *the inherent spatial structure of the variables*

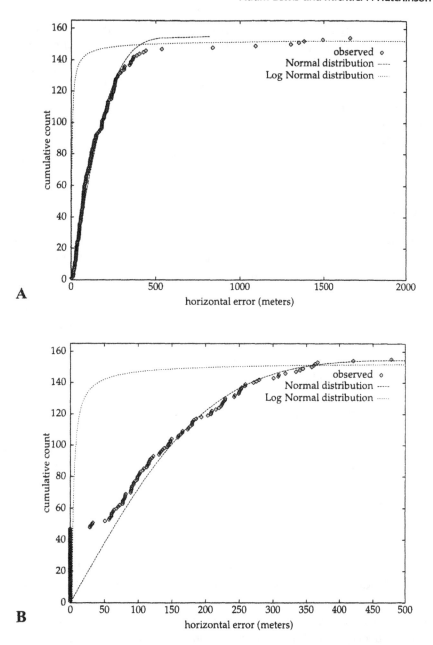

Figure 2.4. Plots of the cumulative distribution of horizontal error for S_1 quadrats (a) and S_2 quadrats (b); and normal and log-normal probability models fitted using the method of moments. An isometric error distribution is assumed. For S_1, the normal distribution was fitted after exclusion of extreme values. Neither probability model deals with S_1 very well; however, the normal distribution was chosen in preference to the log-normal.

being observed. Thus, even the coarse models of spatial error used here are sufficient to quite accurately model the quality of the quadrat data sources for overlay with other spatial data themes.

Table 2.2. Modeled and Observed (in brackets) Correlations (R^2) between Observations of Elevation and Slope from True Data Point Locations and Data Point Locations Containing Spatial Error. (Epsilon = 2σ.)

	Slope (percent rise)	Elevation (meters)
Point locations from S$_1$. Epsilon = 384	**0.303** (0.181)	**0.957** (0.952)
Point locations from S$_2$. Epsilon = 320	**0.343** (0.318)	**0.967** (0.976)

DISCUSSION

These results have wide-ranging practical value. They suggest that the detail of the probability error model $f(\mathbf{x})$, is less important in determining data quality than the inherent properties of the surfaces being interrogated. Only the main features of $f(\mathbf{x})$ seem to be important.

The results also suggest an approach to management of GIS error which integrates spatial and attribute error. If the value of a variable is estimated from interrogation of a data-surface Z, where Z is a model with residual (attribute) error σ^2_{eZ}, the variance of Z, at any given place, due to spatial and attribute error can be calculated as the expected semivariance, plus the variance of the residual error of the model:

$$\sigma^2_Z = E[\gamma_z] + \sigma^2_{eZ} \text{ where } E[\gamma_z] = \int f_z(\mathbf{h}) \; \gamma_z(\mathbf{h}) \; d\mathbf{h} \tag{8}$$

In which $f_z(\mathbf{h})$ is the probability density function of the spatial errors in variable Z, while $\gamma_z(\mathbf{h})$ is the semivariance of the variable Z, readily estimated from the data-surface Z.

Applying Equation 8 to the DEM used here, assuming isotropic normally distributed spatial errors with *epsilon* = 25, and an RMS of 5 (σ^2_{eZ} = 25) gives

$$E[\gamma_z] \approx 3.7, \; \sigma^2_Z = 3.7 + 25$$

Thus (for elevation) the influence of spatial error is small compared with residual error.

A number of assumptions and limitations apply to these findings. The assumption of an isotropic Gaussian distribution of spatial errors is conservative in that anisotropic errors, or error distributions which have a nonzero mean will, according to the model, lead to a more rapid decline in data quality. Although for the data studied here these limitations appear to be of minor consequence, this area is worthy of further attention.

Within this study it has been appropriate, given the small area of the vegetation quadrats (20×20 meters) relative to the elevation model lattice, to assume that these may be treated as point data. In practice larger quadrat areas are not uncommon, in which case the use of semivariance formulas based implicitly on point estimates is inappropriate. Provided the quadrats are isometric, this difficulty can be overcome by first integrating (averaging) the data-surface by a moving window of area corresponding to the area of the sample quadrat. The surface so derived is the data-surface as "seen" through the sample quadrats.

The method is capable of generating an estimate of the R^2 value which can be expected during an overlay operation; however, there are no error-bounds placed on this estimate.

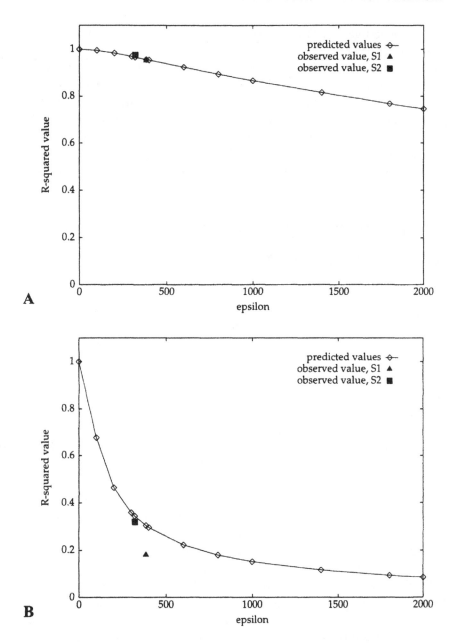

Figure 2.5. (a) Predictions of R^2 value for elevation given a wide range of error bands, compared with observed R^2 values from S_1 and S_2. (b) Predictions of R^2 value for slope given a wide range of error bands, compared with observed R^2 values from S_1 and S_2.

Clearly, uncertainty in the estimate can arise from two sources, incorrect specification of the error model, especially the scaling parameter, and poor estimation of the variogram, which is assumed to be isotropic. The former problem can be addressed to some extent by stating upper and lower bounds within which the positional error parameters would certainly lie, and simulation within this range as illustrated by Figure 2.5, but nonetheless requires further investigation. The isotropy constraint on the variogram is removed in the developments described below.

Assuming that both of these inputs are accurate, we have not pursued the question of the variance of the estimate of the expected R^2 value; i.e., for a given actual sample size, what bounds may be placed on the expected R^2? This may warrant further investigation.

For practitioners, the most serious limitation of these methods is that they apply only to continuous data-surfaces defined on an interval or ratio scale. Thus, these methods cannot, in current form, be applied to the problem of point-in-polygon overlay (although extensions to address this are reasonably straightforward), nor to the problem of polygon-polygon overlay.

FURTHER DEVELOPMENT AND APPLICATION

Implementation in ARCINFO® GRID

In this section we describe further development of these concepts into a tool accessible to GIS manager or analysts using the ARCINFO® GRID software, written in the ARCINFO® macro language (AML). We relax the assumption of isotropy in the variogram model, and improve the error model by representing the error in x and y as independent Gaussian processes.

Accepting the finding that the *application* is in general more critical than the detailed characteristics of the spatial error distribution, we have a simple, widely applicable method for the analysis of point data quality for overlays. Generalization of the method to a two-dimensional variogram improves the accuracy of the method, but also converts the integration implied by Equation 5 to a two-dimensional "spatial" overlay problem, easily implemented in a GIS. This is convenient as it helps to bridge the functional gap between spatial statistical analysis and more commonplace GIS operations.

The Gaussian error distribution models given above can be improved by allowing x and y to follow independent identical Gaussian distributions, from which it follows that the magnitude of the error is modeled as a Weibull distribution (Conradsen and Nielson, 1984), and extending to a two-dimensional form. This distribution may be represented as a regular lattice or set of grid cells of arbitrarily small size. Clearly, the finer the lattice, the more accurate the discrete model of the continuous probability density function becomes; however, a two-dimensional grid-cell representation of a Gaussian bell curve with 100 cells in both the x and y directions, as illustrated in Figure 2.6, is at least sufficiently accurate for our needs. This probability model can be rescaled to any size, corresponding to increasing or decreasing the variance of the distribution, simply by alteration of the grid cell size. In this way we can model the effects of a range of magnitudes of spatial error.

The two-dimensional variogram differs from the more common unidimensional variogram in that the lag is attributed to offsets in both x and y, rather than simply the magnitude of the distance $\sqrt{(dx^2 + dy^2)}$. Thus the semivariance at lag (2,2) is not assumed to be equal to the semivariance at lag (2,-2); however, as the two-dimensional variogram is symmetrical through the origin, lags (2,2) and (-2,-2) are regarded as identical.

For any two-dimensional gridded data set the expected semivariance for any given distance lag can be estimated by a sequence of simple offset and overlay operations. That is, a copy of the data set is taken, offset from the original by the vector (x,y), after which the

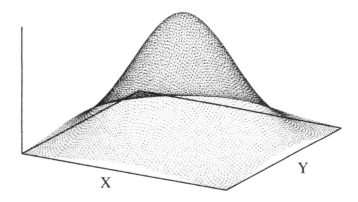

Figure 2.6. A two-dimensional joint probability density function for two independent identically distributed Gaussian variables, *x* and *y*, represented as a 100 * 100 cell grid. The horizontal axes are in arbitrary units of distance, while the vertical dimension is probability density, which is assumed to reach a maximum at the origin, (0,0).

square of the difference between the cell values is calculated for each cell. Half of the mean of this value is the traditional semivariance measure for the lag (*x,y*). This can be converted to an R^2 value by subtraction from, and scaling by, the data set variance. Iteration through a range of lags in the *x* and *y* directions generates a matrix of semivariance values corresponding to these lags. This matrix can be treated as a grid in a GIS in the same way that the probability density function described above can be represented as in Figure 2.7. The two-dimensional variogram surface can be converted to a two-dimensional squared correlogram using Equation 3. Again, this is a straightforward GIS operation.

Multiplication of the squared correlogram surface by the probability density function surface, followed by summation, is the equivalent to the integration of Equation 5. It may be noted that the variogram is observed from the data set, not modeled on the basis of a limited number of samples, so there is no observable nugget variance.

For convenience, and arguably at the expense of execution speed, these steps have been implemented in the ARCINFO® GRID software in a series of standard, portable, AML procedures. sv.aml calculates the two-dimensional variogram for a data set over some range of lags, sv_2d.aml converts these results to a grid, and DQA.aml (Data Quality Assessment) determines the effect of spatial error for a range of expected error scenarios.

Application

In this section we demonstrate the application of DQA.aml to the analysis of a survey intended to determine relationships between an important forest pathogen, *Phytophthora cinnamomi*, and environmental variables derived from digital terrain modeling. Although the example is in this case post hoc, this is not a requirement.

We examined the following question: are the sample points sufficiently accurately located to allow overlay with terrain models, and, for each terrain variable, how much information is lost due to the spatial error of the samples? The analysis demonstrates that the failure to detect some expected relationships can be explained simply by the estimated spatial error associated with the locations of the sample data.

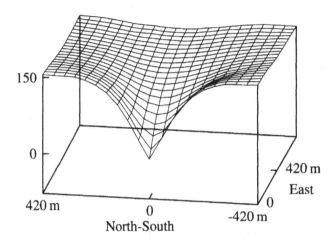

Figure 2.7. A two-dimensional variogram for the variable *SI* (see main text). The variogram is symmetrical through the origin (the lowest point), and extends 420 meters to the north and south and 840 meters to the east. A semivariance of zero (or the nugget variance) occurs at lag (0,0). With increasing lag distance this rises to a sill, at which the semivariance approximates the total variance of the data set (141).

The study area is a 7,600 ha catchment at Anglesea in southeastern Australia (Figure 2.8) with recognized National Estate values, managed under a mining lease (Wilson et al., 1997). *Phytophthora cinnamomi* is widespread in the catchment, and is a significant threat to native vegetation communities. The presence of *Phytophthora* was tested for at 47 sample sites across the catchment.

A DEM, interpolated using ANUDEM (Hutchinson, 1988,1989), to a 20-meter lattice from 1:25,000 scale digital contours with contour interval of 10 meters, was used to model the terrain. Three surfaces were derived from the DEM, guided by a knowledge of the ecology of the pathogen (Marks et al., 1982). Elevation (*ELEV*) was read directly from the DEM. A sun index (*SI*) was defined as:

$$SI = \text{cosine}(aspect) * \text{tangent}(slope)$$

where *aspect* is in degrees (clockwise or anticlockwise) from north, and *slope* is in degrees. The third index was the natural log of the specific catchment area (*LnSCA*).

The specific catchment area (SCA) (Moore et al., 1991), when calculated in a grid-cell GIS, is a measure of the area of the catchment expected to contribute overland flow to each grid cell. Calculation requires a DEM which represents flow directions correctly, and generally proceeds following the algorithms of Jenson and Dominigue (1988). Flow paths are modeled down the catchment, and each grid cell is given a value for the catchment area above it. Progressively higher values of SCA therefore occur from ridges, to downslope areas, drainage lines, streams and rivers, and the variable is therefore presumed to correspond to topographically driven variations in soil moisture, and in the case of *P. cinnamomi*, which is propagated by water movement, to an increased probability of the fungus in the soil. The statistical distribution of SCA is usually highly skewed, and *LnSCA* is therefore the log transform of SCA.

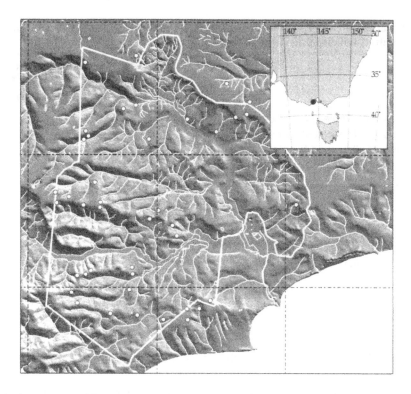

Figure 2.8. Anglesea study area, sample locations and terrain model (20 meter lattice resolution). The overlaid grid-lines are at 5,000 m intervals.

ELEV, SI, and *LnSCA* at each sample site were estimated by overlay of the sample point locations with each surface with using ARCINFO® Latticespot. A number of vegetation measures were also observed at each survey quadrat.

Only two variables, *ELEV* and *SI* were significant in a logistic regression of *Phytopthora* on all of the terrain and vegetation variables, the significance of the correlation being lower for *SI* (p=0.029 *cf.* p=0.004). Given that *Phytophthora* tends to propagate in a downslope direction by zoospores, and is also generally limited by conducive combinations of soil temperature and soil moisture (Marks et al., 1982), the lack of significance of *LnSCA* was unexpected.

The value of *SI* in the regression is readily explained. As the mean temperatures within the catchment are low compared to the optimal requirements of the species (Marks et al., 1982), topographic warming of northern aspects is likely to be a significant factor in the development of *Phytophthora* at a site.

The value of *ELEV* in the regression is not immediately clear. Analysis of the relationships between elevation and temperature within the catchment determined that *ELEV* is not acting as a surrogate for temperature in this case, since the expected elevation-based temperature differential in this catchment is a mere 0.8 degrees. It follows that correlations with *ELEV* are either spurious (which is unlikely given the significance of *ELEV* in the model) or the function of some other elevation-related process, such as the tendency for spread down the catchment. However if downslope spread is indeed important, why was it

Table 2.3. Lineage of Survey Quadrat Coordinate Data, and "Guestimation" of the Error Associated with Each Step in This Lineage.

Step (Lineage)	Standard Deviation, in Meters, of the Error (Informed Guesses). (Units are meters)
Record location of sample using 2-dimensional fixes from a handheld single-channel GPS without postprocessing or allowance for elevation. Spheroid presumably WGS84	30
Record locations onto a notepad (assuming no blunders)	0
Transcribe point locations onto a 1:25,000 map base (assuming no blunders)	25
Digitize point locations	12.5
Overall estimate (summed variances$^{0.5}$):	~40 meters

not detected by the variable *LnSCA*, which aims to model this process much more directly than elevation?

The lineage of the sample point location data is far from ideal due to a lack of planning and experience of field staff. This raises the possibility that the error associated with the location of these samples is causing a loss of information in the overlay of the sample points with the terrain model surfaces. To gain an approximate model for the spatial error associated with the sample point locations, which were digitized in a number of steps, we can use some guesses based on experience (Table 2.3), from which an estimate of $2\sigma \approx 80$ meters is reached. If the errors in the *x* and *y* directions follow independent identical normal distributions, then $2\sigma \approx 80$ implies that 63% of the points can be expected to be within 40 meters of the correct location, and 98% of points within 80 meters. Given the 1:25,000 scale of the data sets, this is poor and could certainly be improved with good procedures; however, field practices in the recording of location are notoriously lacking, especially with historical data, and this level of error is considered realistic. This estimate cannot be verified since the true locations are, in this case, unknown.

Two-dimensional variograms and correlograms for *ELEV, SI,* and *LnSCA* were derived from the data sets using sv.aml and sv_2d.aml. The effect of the spatial error in the sample point data on the outcomes of overlay was modeled using these variograms and a two-dimensional Gaussian distribution, as described above, using DQA.aml.

Results are illustrated in Figure 2.9. Figure 2.9a illustrates that the error of the samples is so small relative to the scales of variation in *ELEV* that there is no loss of information during the overlay ($R^2 = 1$). Figure 2.9b illustrates that for *SI* the correlation expected between *SI* for the true locations versus the observed locations is in the order of $R^2 = 0.61$. Thus, some information is lost through the error of the point locations. Conversely, more accurate sample locations may lead to an increase in the significance of *SI* in the logistic regression model for *Phytophthora*. Figure 2.9c shows that the expected R^2 for *LnSCA* is only 0.08, reflecting the fact that *LnSCA* varies substantially at distances within the range

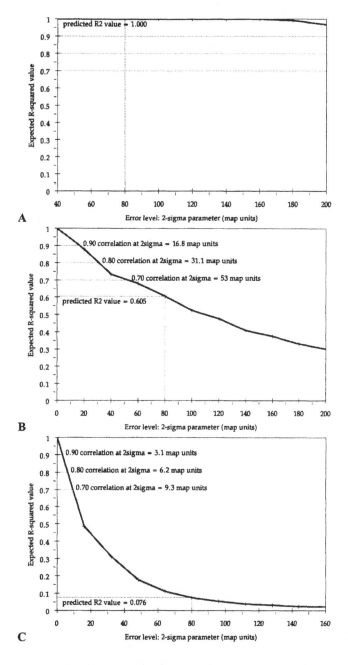

Figure 2.9. Results of the data quality analysis for overlay of sample data points with three derived terrain attributes. Data quality is measured as the expected R^2 value for the attribute, from 1 (ideal; no information is lost) to 0 (the sample locations are so inaccurate that all information is lost). (a) The error in the sample is minor relative to the spatial structure of *ELEV*, and the point data perform perfectly in this analysis (and would do so even if far less accurate). (b) Overlay with *SI*. The model predicts that the error in the locations of the sample points will lead to a loss of 39% of the information (variance) associated with this variable. Improved locations may lead to increased significance of *SI* in the regression model. (c) Overlay with *LnSCA*. The model predicts that virtually all information from this variable is lost in the overlay process as a result of the spatial error of the point data. Furthermore, very accurate locations would be required to avoid loss of information.

of the error expected in the sample, as a result of which the sample accuracy is inadequate to capture information on catchment areas. Even if there was a strong relationship between *LnSCA* and *Phytophthora,* we would not be able to detect this relationship from the available data.

This brief analysis demonstrates that spatial accuracy may be an important consideration when samples are used with GIS data, and that the sample design stage can and should include an assessment of the required spatial accuracy of the sample locations.

CONCLUSION

We have shown that at least in some cases the fitness for use of a data set can be assessed, and predicted, quantitatively. When a set of points is overlaid with numerical data-surfaces such as elevation models, the expected R^2 value between the outcome of the overlay with and without positional error offers a simple and meaningful measure of the suitability of the point data for the overlay operation. We have also shown that this suitability tends to be determined at least as much by the use of the data as by the fine details of the positional error in the point data; this is in agreement with our expectations.

Furthermore, our results demonstrate that the expected R^2 value can be predicted with minimal knowledge of the details of the positional error of the points used in the overlay; a crude estimate, used as the variance parameter of a simple error model, appears to suffice. These predictions can be made using standard GIS tools which have been implemented as ARCINFO® AML procedures available from the first author.

REFERENCES

Allen T.F.H. and T.B. Starr. *Hierarchy: Perspectives for Ecological Complexity.* University of Chicago Press, 1982.

Bolstad, P.V., P. Gessler, and T.M. Lillesand. Positional uncertainty in manually digitised map data. *Int. J. Geogr. Inf. Syst.,* 4, 399, 1990.

Burrough, P.A. *Principles of Geographical Information Systems for Land Resources Assessment.* Clarendon Press, Oxford, 1986.

Chrisman, N.R. *Exploring Geographical Information Systems.* John Wiley & Sons, Brisbane, 1997.

Conradsen, K. and L.B. Nielson. Review of Weibull statistics for estimation of wind speed distributions. *J. Climate Appl. Climatol.,* 23, pp. 1173–1183, 1984.

Goodchild, M.F. The Issue of Accuracy in Global Databases, in *Building Databases for Global Science. Proceedings of the First Meeting of the International Geographical Union Global Database Planning Project,* H. Mounsey and R.F. Tomlinson, Eds., Tylney Hall, Hampshire, UK, May 1988. Taylor and Francis. London & New York, 1988.

Goodchild, M.F., S. Guoqing, and Y. Shiren. Development and test of an error model for categorical data. *Int. J. Geogr. Inf. Syst.,* 6, pp. 87–104, 1992.

Hutchinson, M.F. Calculation of hydrologically sound digital elevation models. *Proceedings of the Third International Symposium on Spatial Data Handling.* Sydney, Australia, 1988.

Hutchinson, M.F. A new procedure for gridding elevation and stream line data with automatic removal of spurious pits. *J. Hydrol.,* 106, pp. 211–232, 1989.

Jenson, S.K. and J.O. Dominigue. Extracting topographic structure from digital elevation data for geographic information system analysis. *Photogrammetric Eng. Remote Sensing,* 54, pp. 1593–1600, 1988.

Kiiveri, H.T. Assessing, representing and transmitting positional uncertainty in maps, *Int. J. Geogr. Inf. Sci.,* 11, pp. 33–52, 1997.

Marks, G.C., B.A. Furhrer, and E.M. Walters. *Tree Diseases in Victoria.* Forests Commission of Victoria, Melbourne, Australia, 1982.

Moore, I.W., R.B. Grayson, and A.R. Ladson. Digital terrain modelling: A review of hydrological, geomorphological, and biological applications. *Hydrol. Process.,* 5, 1991.

Panofsky, H.A. and J.A. Dutton. *Atmospheric Turbulence.* John Wiley & Sons, New York, 1984.

Turner, L.A. and S.G. Mueck. *The Vegetation of the Sardine, Rich and Ellery forest blocks, Orbost region, Victoria.* Silvicultural Systems Project Technical Report Number 9, 1992. Department of Conservation and Environment, Melbourne, Australia.

Wilson, B.A., A. Lewis, and J. Aberton. *Conservation of National Estate Communities threatened by Cinnamon Fungus at Anglesea, Victoria.* Report to the Department of Natural Resources and Environment, Victoria, 1997.

Chapter 3

Generalized Linear Mixed Models for Analyzing Error in a Satellite-Based Vegetation Map of Utah

Gretchen G. Moisen, D. Richard Cutler, and Thomas C. Edwards

INTRODUCTION

Thematic accuracy of vegetation cover maps derived from satellite imagery may be related to many factors, including elevation, aspect, slope, local heterogeneity, and distance to vegetation boundaries. Exploring the relationship between the components of the vegetation classification model and its uncertainty is a logical step in an analysis of map error that is sensitive to both map use and subsequent improvements of the map.

Although many new techniques are being explored to address map uncertainty, generalized linear mixed models (GLMMs) have yet to be applied. Through a GLMM, data from any one of a variety of continuous and discrete distributions can be linked to a linear structure that may contain both fixed and random effects. GLMMs can also account for correlation among observations as well as among random effects terms in the linear structure (Wolfinger and O'Connell, 1993). This flexibility may prove valuable in addressing map uncertainty, as well as have numerous other broadscale applications.

In this study, we use a GLMM to explore the relationship between the error in the blackbrush cover-type of a vegetation map of Utah and various topographical and heterogeneity components of that map.

DATA

A cover-map of Utah, ~219,000 km^2 in size, was developed from a statewide Landsat Thematic Mapper (TM) mosaic created from 24 scenes at 30 m resolution (Homer et al., 1997). A total of 38 cover-types were modeled. Modeling was accomplished using a four-step modeling approach. Steps included: (1) the creation of a statewide seamless mosaic of TM images; (2) the subsetting of the mosaic into three ecoregions, the Basin and Range, Wasatch-Uinta, and Colorado Plateau (after Omernik, 1987); (3) the association of 1,758 statewide field training sites to spectral classes; and (4) the use of ecological parameters based on elevation, slope, aspect, and location to further refine spectral classes representing multiple cover-types.

Following development of this cover-map, field data were collected to assess its thematic accuracy. Of primary interest were estimates of by-class and by-ecoregion accuracy of the map at the base model of one ha. A total of 100 7.5-min quadrangles were randomly selected approximately proportional to the area of the three ecoregions. Because of the increased expense in accessing remote areas, two strata were identified on each quadrangle based on proximity to roads. The "road" stratum consisted of all land within 1 km of a secondary or better road. All other lands fell within the "off-road" stratum. On each quadrangle, 10 points were randomly selected within the road stratum, while 10 were collected in a randomly oriented heterogeneous linear cluster within the off-road stratum. This design was developed following an investigation of intracluster correlation and a cost function on the relative efficiency of cluster sampling in a statewide map (Moisen et al., 1994). More details on the sample design and analysis of map accuracy is reported in Edwards et al. (1998).

For this study, we focused on modeling map error within the blackbrush cover-type of the Colorado Plateau region in Utah. Blackbrush is an important source of forage for many species of wildlife. This subset of the statewide data consisted of 96 sample points collected on two strata (road, off-road) on 15 quadrangles. Anywhere from 1 to 10 blackbrush points were available for each quad/stratum combination. Clustered blackbrush data in the off-road stratum were not necessarily adjacent sample points because blackbrush polygons were often intermixed with other cover-types not considered in this analysis.

MODEL

Using a logit link function, the binary response (correctly classified / incorrectly classified) was modeled as a function of both fixed and random effects while accounting for several covariance structures for random effects and for spatially autocorrelated errors. For this analysis, the observations on the 96 sample points were coded as 1 when the mapped cover-type agreed with the ground cover-type, and as 0 when they did not agree. Following the notation of Wolfinger and O'Connell (1993), define y to be our data vector of 96 0s and 1s satisfying

$$\mathbf{y} = \mu + \varepsilon \tag{1}$$

We used a logit link function

$$g(\mu) = \log\{\mu / (1 - \mu)\} \tag{2}$$

and modeled

$$g(\mu) = \mathbf{X}\beta + \mathbf{Z}\nu \tag{3}$$

Here, β is a vector of unknown fixed effects with known model matrix $\mathbf{X}$, and ν is a vector of unknown random effects with known model matrix $\mathbf{Z}$. Assume $E(\nu) = \mathbf{0}$ and $\text{cov}(\nu) = \mathbf{G}$, where $\mathbf{G}$ is unknown. An effect may be considered fixed if the inference space is limited to the observed levels of that effect. An effect may be considered random if the inference space is applied to a population of levels, not all of which are observed. Fixed effects considered in this application included both discrete and continuous variables. Because quadrangle maps were randomly selected for subsampling from a population of quadrangles, quadrangles were modeled as random effects. Also, ε is a vector of unobserved errors with $E(\varepsilon|\mu) = \mathbf{0}$ and

$$\text{cov}(\varepsilon|\mu) = \mathbf{R}_\mu^{1/2} \mathbf{R} \mathbf{R}_\mu^{1/2} \tag{4}$$

Here $\mathbf{R}_\mu$ is a diagonal matrix containing evaluations at μ of the variance function

$$V(\mu) = \mu(1-\mu) \tag{5}$$

$\mathbf{R}$ and $\mathbf{G}$ were modeled using covariance structures detailed below.

Fixed Effects

Nine fixed effects variables were considered. Three were topographical variables used in the classification model itself. These were extracted from a 90 m Digital Elevation Model and include elevation in meters (ELEV), slope in degrees (SLOPE), and aspect in degrees. A radiation index (TRASP), used by Roberts and Cooper (1989), takes the form

$$\text{TRASP} = \frac{1 - \cos\big((\pi/180)(\text{aspect} - 30)\big)}{2} \tag{6}$$

This transformation assigns a value of zero to land oriented in a north-northeast direction, the coolest and wettest orientation in Utah, and a value of one on the hotter, drier south-southwesterly slopes.

In addition to the three topographical variables, we considered four different measures of heterogeneity surrounding the sample point. Richness (RICH) is defined as the number of cover-types found in the surrounding eight pixels. The other three heterogeneity variables, evenness (EVEN) and two measures of diversity (D_1 and D_2), are defined in Table 3.1. Higher values for all indices indicate increasing heterogeneity.

Two other fixed effects considered were the minimum distance in meters to a different map cover-type (DIST) and a variable indicating membership in the road or off-road stratum (STRATA). Strata was the only categorical fixed effect variable. All others were continuous.

Covariance Structures

Because quadrangle maps (QUAD) were randomly selected for subsampling from the statewide population of quadrangles, these were included as random effects in the GLMM. Three covariance models for $\mathbf{G}$ were considered (Table 3.2).

A spherical spatial covariance structure, illustrated in the last row of Table 3.2, was considered for $\mathbf{R}$. Here covariance between sample points is modeled as a function of dis-

Table 3.1. Measures of Heterogeneity. (Here S equals richness, n equals 8 pixels, and n_i is the number of pixels belonging to the ith of S cover-types.)

Variable	Formula	(Eq.)	Reference
D1	$D_1 = \exp\left(-\sum_{i=1}^{S}\left[\left(\frac{n_i}{n}\right)\ln\left(\frac{n_i}{n}\right)\right]\right)$	(7)	Hill, 1973
D2	$D_2 = \left[\sum_{i=1}^{S}\frac{n_i(n_i-1)}{n(n-1)}\right]^{-1}$	(8)	Simpson, 1949
EVEN	$\text{EVEN} = (D_2-1)/(D_1-1)$	(9)	Ludwig and Reynolds, 1988

Table 3.2. Covariance Structures for G and R.

Structure	Form	(Eq.)
Simple	$G_{ij} = \sigma^2$ for $(i=j)$, else 0	(10)
Compound symmetry	$G_{ij} = \sigma_1^2 + \sigma^2$ for $(i=j)$, else σ_1^2	(11)
Varying coefficients	$G_{ij} = \sigma_{ij}^2$ for $(i=j)$, else 0	(12)
Spherical spatial	$R_{ij} = \sigma_e^2\left[1-\left(3d_{ij}/2\rho\right)+\left(d_{ij}^3/2\rho^3\right)\right]$ for $d_{ij} \leq \rho$, else 0	(13)

tance between those points, accounting for both correlation between clustered locations and potential correlation between sample points in relatively close quadrangles. Although numerous spatial structures could have been tried, the spherical structure is very flexible and converged more readily in preliminary trials.

Model Fitting Strategy

Parameters in the GLMM were estimated through pseudo-likelihood procedures as described in Wolfinger and O'Connell (1993) using a SAS macro supplied by Russ Wolfinger of SAS Institute Inc. This macro uses PROC MIXED and the Output Delivery System, requiring SAS/STAT and SAS/IML release 6.08 or later.

Model fitting proceeded as follows. After identifying quadrangle maps as our random effects, all fixed effects were included in the model. We tried all covariance structures for **G** as listed in Table 3.2 along with the spatial covariance structure for **R**. The best covariance structure for **G** was selected based on Akaike's Information Criterion and Schwarz's Bayesian Criterion (Wolfinger, 1993). Having chosen suitable covariance structures, fixed effects were then dropped based on significance of parameter estimates, likelihood ratio tests, and predictive capability of the model. Because of collinearity, the four measures of heterogeneity were considered in the model separately.

Table 3.3. Parameter Estimates and Their Standard Errors for Final GLMM.

Parameter	Estimate	SE	$Pr > \chi^2$ or $(Z*)$
Intercept	−1.78	0.85	0.04
STRATA (off-road)	−1.39	0.59	0.02
SLOPE	0.12	0.07	0.08
D_2	0.58	0.31	0.06
ρ	322.46	124.86	0.01*
s^2	1.63	1.37	0.23*

RESULTS

Likelihood ratio tests and parameter significance levels led us to favor a parsimonious model containing only STRATA, SLOPE, and D2 as fixed effects. Exclusion of other variables had little impact on the predictive capability of the model based on confusion matrices and plots of predicted values from different model trials. The covariance structures selected were simple and spherical for **G** and **R**, respectively. The signs of the fixed effects parameters indicate the relationship between error and the variables (Table 3.3). In this case, positive values for SLOPE and D_2 suggest that error increases as SLOPE and D_2 increase (Figures 3.1a–b). In contrast, the negative value for the off-road stratum indicates that probability of error is less in the off-road stratum and greater in the road stratum. The signs and magnitudes of the fixed effects parameters shown in Table 3.3, however, should be interpreted with caution unless potential problems with multicollinearity and the possibility of interaction terms have been examined. The estimate of ρ, the spatial covariance parameter in **R**, suggests that spatial dependence is negligible between sample units greater than 322 meters apart (Figure 3.1c).

DISCUSSION

Frequently in broadscale sample surveys, clustered and subsampling sample designs are adopted for the sake of efficiency in estimation of population means and totals. However, such cost-effective designs can hamper efforts to further explore ecological relationships under a classical linear model framework by violating model assumptions like independence and normality of errors. In this study we illustrated how a GLMM affords the flexibility to analyze sample survey accuracy data. Through a GLMM we were able to include both fixed and random effects, account for spatially autocorrelated errors, and allow for a variety of covariance structures for the random effects.

The model fitted for blackbrush contained some information for map users beyond simple probability of misclassification by cover-type, and also provided information to the map-maker for model improvements. We learned that incorrect mapping of this cover-type tended to occur near roads in steep and heterogeneous areas. The fact that strata was a significant contributor to our model of map error could be an indication that vegetation away from roads differs from that near roads, and is governed by environmental factors not accounted for in the other fixed effects. However, the better performance in the off-road stratum could be an artifact of different quality in data collection efforts between the two strata. For

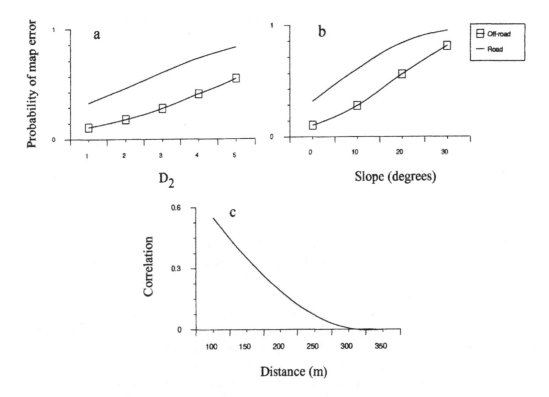

Figure 3.1. a–b: Relationship between probability of map error and fixed effects. 1c. Correlation modeled as a function of distance between sample points.

example, georeferencing data in linear clusters may have been easier or done with greater care, making off-road data less subject to positional error.

The inclusion of slope in our model might highlight the difficulty of classification in steep and often shadowy areas. Slope was not included in the initial classification model for blackbrush in the Colorado Plateau, and its inclusion might improve maps of that cover-type. The notion that classification in heterogeneous areas is tougher than in homogenous areas is not new, but our model helps determine the magnitude of heterogeneity's contribution to error. Also, numerous indices of heterogeneity are available in the literature and we illustrated that some indices may be making a more significant contribution than others to models of map error. Map users might also find the model results helpful, making them more skeptical of mapped blackbrush on steeper slopes and in more heterogeneous areas.

ACKNOWLEDGMENTS

We would like to thank our reviewers for their thoughtful comments. We are also grateful to Ron Tymcio who tackled many GIS challenges, David Early for data collection, Collin Homer who served as our patient and ever-present practical conscience, and especially Scott Bassett, who can program anything, fast. This chapter is a collaborative effort between the USDI National Biological Service's Gap Analysis Program and the USDA Forest Service Intermountain Research Station, Ogden, Utah.

REFERENCES

Edwards, T.C., Jr., G.G. Moisen, and D.R. Cutler. Assessing map accuracy in an ecoregion-scale, remotely-sensed cover map. *Remote Sensing Environ.,* 63, pp. 73–83, 1998.

Hill, M.O. Diversity and evenness: A unifying notation and its consequences. *Ecology.* 54, pp. 427–432, 1973.

Homer, C., R.D. Ramsey, T.C. Edwards, Jr., and A. Falconer. Landscape cover-type modelling using a multi-scene TM mosaic. *Photogrammetric Eng. Remote Sensing.* 63, pp. 59–67, 1997.

Ludwig, J.A. and J.F. Reynolds. *Statistical Ecology.* John Wiley & Sons, New York, 1988.

Moisen, G.G., T.C. Edwards, Jr., and D.R. Cutler. Spatial Sampling to Assess Classification Accuracy of Remotely Sensed Data, in *Environmental Information Management and Analysis: Ecosystem to Global Scales.* Taylor and Francis, London, 1994, pp. 159–176.

Omernik, J.M. Map supplement: Ecoregions of the conterminous United States. *Ann. Assoc. Am. Geogr.* 77, pp. 118–125 (map), 1987.

Roberts, D.W. and S.V. Cooper. Concepts and Techniques of Vegetation Mapping, in *Land Classifications Based on Vegetation: Applications for Resource Management.* USDA Forest Service General Technical Report INT-257, Ogden, UT, 1989, pp. 90–96.

Simpson, E.H. Measurement of diversity. *Nature.* pp. 163, p. 688, 1949.

Wolfinger, R.D. Covariance structure selection in general mixed models. *Commun. Statistics: Simulation Computation.* 22, pp. 1079–1106, 1993.

Wolfinger, R.D. and M. O'Connell. Generalized linear mixed models: A pseudo-likelihood approach. *J. Stat. Computation Simulation.* 48, pp. 233–243, 1993.

Chapter 4

Increasing Spatial Precision and Accuracy for Monitoring Peatlands in Switzerland by Remote Sensing Techniques

Michael Köhl, Hans Jörg Schnellbächer, and Andreas Grünig

INTRODUCTION

In 1984 and 1990, respectively, the Swiss national bog and fenland inventories were carried out (Grünig et al., 1986; Broggi, 1990). The area of Swiss mire habitats identified by these two inventories totals approximately 30,000 ha, which is less than 1% of the total area of Switzerland. These inventories included approximately 1,700 mires with a total of 20,000 ha, which are regarded to be "Sites of National Importance" (Grünig, 1994; Küttel, 1995). As by federal legislation those sites have to be protected and monitored, a concept for monitoring national peatlands to meet the legal requirements had to be developed. The focus of the monitoring program is to obtain results for the entirety of Swiss peatlands and to describe the situation of individual sites (objects).

Long-term monitoring of Swiss peatlands has to cover two different objectives: (1) to detect changes in the total area of peatlands, and (2) to detect changes of the area covered by distinct vegetation types. As vegetation cover is an indicator for the quality of site conditions, it serves the purpose of monitoring site changes as well. The methods applied to meet the first objective have been described elsewhere (Kleinn, 1991; Köhl and Kleinn, 1994; Köhl, 1994) and are not subjects of this chapter.

Two monitoring approaches will be presented: surveys based on systematically distributed sample plots, and a two-phase sampling design based on the combination of data from aerial photographs and field assessment. The accuracy (bias) of both methods as well as the precision for estimating current state and change is investigated by simulation studies and pilot surveys on a test site.

SAMPLING ALTERNATIVES

A traditional approach in vegetation surveys is to establish and assess a set of systematically distributed sampling units and remeasure it at successive occasions. In forest surveys this design is known as control sampling (Schmid, 1967) or continuous forest inventory

(Stott, 1947). On each plot either the percentage of area cover by species or the presence or absence of a species can be recorded. Preliminary studies have shown that the assessment of the percentage area cover on sample plots is prone to subjective observer bias, which cannot be reduced by sample size (see Gertner and Köhl, 1992, for details). Thus we will concentrate on the second approach, which is more robust in respect to observer bias.

The proportion of the reference area covered by a species can be estimated by defining a variable y, which can take the value 1 if the species is observed on the plot and the value 0 in all other cases. For large populations the proportion $\hat{P}$ can be estimated as follows:

$$\hat{\overline{Y}} = \frac{\sum y_i}{n} = \frac{a}{n} = \hat{P} \tag{1}$$

with variance

$$\hat{S}^2 = \frac{\sum \left(y_i - \hat{\overline{Y}} \right)^2}{n-1} = \frac{n}{n-1} \hat{P}(1-\hat{P}) \tag{2}$$

where n is the number of sampling points and y_i is the observation on plot i. The variance of $\hat{P}$ is estimated by

$$v\left(\hat{P} \right) = \frac{\hat{P}(1-\hat{P})}{n-1} \tag{3}$$

If the survey is based on remeasurement of the same set of permanent plots, change between two successive occasions, $\overline{C}$, and its variance $v(\overline{C})$, can be calculated by taking the difference of the proportions $\hat{P}_i$ on each occasion.

$$\hat{\overline{C}} = \hat{\overline{Y}}_2 - \hat{\overline{Y}}_1 = \hat{P}_2 - \hat{P}_1 \tag{4}$$

$$v(\hat{\overline{C}}) = var(\hat{P}_1) + var(\hat{P}_2) - 2\, r_{yx} \sqrt{var(\hat{P}_1)} \sqrt{var(\hat{P}_2)} \tag{5}$$

where $\hat{\overline{Y}}_i$ = $\hat{P}_i$ = proportion on occasion i (i=1,2)
r_{yx} = correlation coefficient of observations on occasion 2 and occasion 1

In forest survey applications it has often been shown that cost-efficiency can be improved by combining information assessed in aerial photographs and field assessments. A two-phase (or double) sampling for stratification design has been introduced in the second Swiss national forest inventory (Köhl, 1994), where in a first phase strata weights are assessed on a large sample of aerial photographic plots and variables of interest are assessed on a smaller sample of field plots. This design is proposed as an alternative to monitoring the area cover percentage in the Swiss peatland monitoring program.

Applying a double-sampling for stratification design, the mean proportion of area cover of a plant species, $\hat{\bar{Y}}_{ds}$ and its variance $v(\hat{\bar{Y}}_{ds})$ is estimated as follows:

$$\hat{\bar{Y}}_{ds} = \sum_{h=1}^{L} \frac{n_h'}{n'} \hat{\bar{Y}}_h = \hat{P} \tag{6}$$

$$v\left(\hat{\bar{Y}}_{ds}\right) = \sum_{h=1}^{L} \frac{n_h'-1}{n'-1} \frac{n_h'}{n'} v(\hat{\bar{Y}}_h) + \sum_{h=1}^{L} \frac{1}{n'-1} \frac{n_h'}{n'} \left(\hat{\bar{Y}}_h - \hat{\bar{Y}}_{ds}\right)^2 \tag{7}$$

where

$\hat{\bar{Y}}_h$ = mean value in stratum h, h=1,...L

$v(\hat{\bar{Y}}_h)$ = variance of $\hat{\bar{Y}}_h$, h=1,...L

n_h' = number of photo plots in stratum h, h=1,...L

n' = total number of photo plots = $\sum n_h'$

L = number of strata

DATA

Among the peatlands covered by pilot surveys, a mire called "Grossmoos" in the Schwändital, Canton of Glarus, has been selected to provide basic data on plant communities and their spatial distribution for a simulation study. Aerial photographs (CIR, 1:5,000) of the "Grossmoos" were taken in 1994. The aerial photographs were scanned and analyzed by means of digital photogrammetry using a HELAVA DPW 770 station. The borderline of the "Grossmoos" was plotted and patches within the total area were assigned and delineated according to differences in texture and color. Similarities of texture and color were used to form nine strata, which can be regarded to reflect distinct plant communities. Thus the construction of strata was not founded on the occurrence of vegetation types or plant communities observed on the ground, but exclusively on attributes (texture and color) assessed in aerial photographs. Each of the delineated patches was grouped into one of the strata and the results of the photo-interpretation were printed in mapped form. Based on a preliminary map, a vegetation survey was conducted. For the application of a two-phase design, however, only strata have to be assigned to photo plots, but no map is required. For each patch a field assessment was conducted and the plant species observed and their cover recorded. The cover of each plant species within each patch, p, was assessed in five classes:

- not present	$- 0\% < p \leq 0.1\%$
$- 0.1\% < p \leq 1\%$	$- 1\% < p \leq 10\%$
$- > 10\%$	

The results of the vegetation survey were later used to assign vegetation types to the strata formed by aerial photo-interpretation. For the simulation study six plant species reflecting different spatial distributions and abundance classes were selected. Table 4.1 summarizes the spatial distribution and the proportion of area cover for the selected species.

Table 4.1. Selected Species and Corresponding Spatial Distribution and Area Cover of Patches, in Which Species Have Been Observed.

Species	Spatial Distribution of Patches	Proportion of Area Cover within Patches
Andromeda polifolia	present on contagious part of the area	all classes
Anthoxantum odoratum	clustered	frequent, majority >0.1%
Scheuchzeria palustris	clustered in small parts of the area	frequent, majority >0.1%
Senecio alpinus	clustered at the margin	rare, majority <1%
Tofieldia calyculata	present on small contagious part of the area	frequent, majority >0.1%
Vaccinium myrtillus	present on majority of patches	frequent, majority >0.1%

Table 4.2. Design Alternatives Applied in the Simulation Study.

Sampling Design	Grid Density	Sample Size
1-phase sampling	5 * 5 m	6369
	10 * 10 m	1599
	20 * 20 m	402
2-phase sampling for stratification	1st phase[a]: 5 * 5 m	1st phase[a]: 6369
	2nd phase[b]: 10 * 10 m	2nd phase[b]: 1599
	1st phase[a]: 10 * 10 m	1st phase[a]: 1599
	2nd phase[b]: 20 * 20 m	2nd phase[b]: 402

[a] Aerial photography.
[b] Field assessment.

SIMULATION STUDY

In a Monte Carlo simulation study 1-phase and 2-phase sampling designs were tested. The field plots as well as the photo plots were systematically distributed over the entire area. In two phase designs the photo plots were used to estimate the stratum weights. On field plots the presence or absence of each of the six species was recorded. Table 4.2 summarizes the sampling designs and grid densities included in the simulation study.

The simulation study was based on the map obtained by the aerial photo-interpretation and the vegetation survey. In the simulation study an exact proportion of area cover within the observed cover class was randomly selected and assigned to each patch. To simulate the spatial distribution within each patch a matrix of pixels with 2.5*2.5 m spatial resolution was laid over the entire area of the "Gross Moos." For each pixel a random number was drawn and compared to the proportion of area coverage of a species within the patch, in which the pixel is located. If the random number was smaller than the randomly assigned proportion, the species under concern was assumed to be present on the plot, in all other cases the species was assumed not to be present on the pixel. To test the efficiency of the five design alternatives for the estimation of change, the area cover of each species was altered by 10%. Thus the absolute value of changes of percent area cover was higher for frequent species than for species with low percent area cover. Based on the randomly gen-

Table 4.3. *Scheuchzeria palustris* (P_{true} = 1.1416, C_{true} = 0.1041).

				Grid Size (m)		
Field:		5*5	10*10	20*20	10*10	20*20
Photo:					5*5	10*10
Current state	$\hat{P}$	1.2199	1.0175	1.0445	1.0349	1.1521
	bias	0.0782	0.1241	0.0971	0.1067	0.0105
	$s(\hat{P})$	0.1371	0.2493	0.4970	0.0015	0.0023
	rmse	0.1711	0.3157	0.6078	0.1564	0.3104
Changes	$\hat{C}$	0.1114	0.0939	0.0970	0.0956	0.1067
	bias	0.0073	0.0101	0.0070	0.0085	0.0027
	$s(\hat{C})$	0.0409	0.0649	0.0883	0.0001	0.0002
	rmse	0.0525	0.1003	0.1768	0.0616	0.1389

erated maps a second map was produced, in which those pixels were randomly selected, where the presence or absence of species changed. This procedure was repeated for each species. The original stratification with three strata was retained for all maps.

New maps for current state and changes were generated for each of the 1,000 iterations and the five sampling alternatives were simulated in each map. For each sampling alternative, each species, and each iteration the following parameters were estimated:

- current state of percent area cover and its variance using the equations presented above,
- the bias, where bias = $| P_{true} - \hat{P} |$, and
- the root mean square error RMSE, where $\sqrt{v(\hat{P}) + bias^2}$

According to Cochran (1977) the RMSE can be used as a criterion of the accuracy of an estimator.

RESULTS

In Table 4.3 to Table 4.8 the results for the six species are presented. For each of the five design alternatives the estimates proportion, $\hat{P}$, the bias, the standard error of $\hat{P}$, $s(\hat{P})$, and the root mean square error, rmse, are given.

The standard error of ($\hat{P}$) is much smaller for the 2-phase sampling designs than for the 1-phase designs. This holds for both, current values and change, and can be observed independent of the spatial distribution and the frequency of the species. Even for the denser grid size the variance of the 1-phase designs is larger than for the coarsed grids of the 2-phase design. For the estimation of current state and change the bias is smaller for the 2-phase designs with one exception (10*10/5*5 m grid for vaccinium myrtillus). The size of the bias of the 1-phase and 2-phase designs is similar, when the terrestrial grid sizes of the 1-phase designs and the grid size of the photo phase in 2-phase designs are equal. The same effect can be observed for the root mean square error. The studied 2-phase designs arrive at the same root mean square with one-quarter of the field plots of a 1-phase design. As the ratio of change was the same in all strata, an improvement is likely for situations where the changes of cover vary in individual strata.

Table 4.4. *Vaccinium myrtillus* (P_{true} = 8.3311, C_{true} = 0.8335).

		Grid Size [m]				
Field:		5*5	10*10	20*20	10*10	20*20
Photo:					5*5	10*10
Current state	$\hat{P}$	8.3827	8.2727	8.1731	8.1376	8.1916
	bias	0.0516	0.0584	0.1580	0.1935	0.1395
	$s(\hat{P})$	0.3466	0.6875	1.3625	0.0173	0.0234
	rmse	0.3972	0.8147	1.6503	0.3973	0.7934
Change	$\hat{\bar{C}}$	0.8393	0.8291	0.8386	0.8153	0.8386
	bias	0.0057	0.0045	0.0050	0.0183	0.0050
	$s(\hat{\bar{C}})$	0.1139	0.2243	0.4352	0.0022	0.0021
	rmse	0.1442	0.2993	0.5890	0.1707	0.3398

Table 4.5. *Andromeda polifolia* (P_{true} = 0.5267, C_{true} = 0.0501).

		Grid Size [m]				
Field:		5*5	10*10	20*20	10*10	20*20
Photo:					5*5	10*10
Current state	$\hat{P}$	0.5353	0.5353	0.7007	0.5348	0.6274
	bias	0.0086	0.0085	0.1740	0.0081	0.1007
	$s(\hat{P})$	0.0908	0.1795	0.3975	0.0045	0.0012
	rmse	0.1077	0.2237	0.5165	0.1145	0.2494
Change	$\hat{\bar{C}}$	0.0496	0.0470	0.0622	0.0469	0.0564
	bias	0.0005	0.0031	0.0121	0.0031	0.0063
	$s(\hat{\bar{C}})$	0.0263	0.0382	0.0577	0.0002	0.0001
	rmse	0.0343	0.0671	0.1147	0.0425	0.0821

Table 4.6. *Tofieldia calyculata* (P_{true} = 0.0962, C_{true} = 0.0094).

		Grid Size [m]				
Field:		5*5	10*10	20*20	10*10	20*20
Photo:					5*5	10*10
Current state	$\hat{P}$	0.0954	0.1094	0.1418	0.1073	0.1482
	bias	0.0008	0.0133	0.0456	0.0112	0.0521
	$s(\hat{P})$	0.0378	0.0720	0.1214	0.0007	0.0012
	rmse	0.0491	0.1059	0.2116	0.0623	0.1583
Change	$\hat{\bar{C}}$	0.0097	0.0124	0.0172	0.0122	0.0180
	bias	0.0003	0.0031	0.0078	0.0028	0.0087
	$s(\hat{\bar{C}})$	0.0082	0.0117	0.0170	0.0001	0.0002
	rmse	0.0134	0.0227	0.0320	0.0170	0.0256

CONCLUSIONS

The results of the simulation study showed that the same root mean square error can be achieved by 2-phase sampling designs, which use a large set of photo plots to estimate

Table 4.7. *Anthoxantum odoratum* (P$_{true}$ = 3.0541, C$_{true}$ = –0.3047).

Field: Photo:		5*5	10*10	20*20	10*10 5*5	20*20 10*10
				Grid Size [m]		
Current state	$\hat{P}$	3.0558	3.0882	3.7194	3.0964	3.7019
	bias	0.0017	0.0341	0.6653	0.0423	0.6479
	s($\hat{P}$)	0.2151	0.4312	0.9394	0.0038	0.0081
	rmse	0.2470	0.5096	1.2524	0.2313	0.7371
Change	$\hat{C}$	0.3082	0.3091	0.3776	0.3098	0.3756
	bias	0.0035	0.0044	0.0729	0.0051	0.0709
	s($\hat{C}$)	0.0689	0.1352	0.2646	0.0004	0.0007
	rmse	0.0878	0.1779	0.3856	0.0986	0.2282

Table 4.8. *Senecio alpinus* (P$_{true}$ = 1.9300, C$_{true}$ = 0.1767).

Field: Photo:		5*5	10*10	20*20	10*10 5*5	20*20 10*10
				Grid Size [m]		
Current state	$\hat{P}$	1.9203	1.9548	2.1542	1.9571	2.2573
	bias	0.0097	0.0248	0.2242	0.0271	0.3273
	s($\hat{P}$)	0.1717	0.3451	0.7191	0.0019	0.0045
	rmse	0.1947	0.4011	0.8690	0.1730	0.4832
Change	$\hat{C}$	0.1767	0.1828	0.2119	0.1830	0.2227
	bias	0.0000	0.0061	0.0352	0.0063	0.0459
	s($\hat{C}$)	0.0520	0.1009	0.1713	0.0002	0.0004
	rmse	0.0661	0.1374	0.2818	0.0788	0.1891

strata sizes and a subsample of field plots, with about one-quarter of the field plots of a comparable 1-phase design. The efficiency of a 2-phase design proved to be superior.

However, the optimal design can only be selected when cost-efficiency is involved. The optimal design among the design alternative is the one which either leads to the smallest variance for a given cost or which results in the smallest cost for a desired variance. For a desired variance, the total cost of obtaining aerial photographs, aerial photo interpretation, and amortization of photo interpretation equipment should be less than 75% of the total cost of a strictly field-based assessment, referring to the results obtained in the simulation study. A 2-phase design would be superior in terms of cost-efficiency if the cost ratio of field and photo plots would be 1.33:1. Experience has shown that the cost ratio is much better referring to photo samples. Taking into account other side effects of aerial photography such as the documentation of a historical situation, providing a piece of evidence for later investigations, and allowing retrospective analysis at a later date, the superiority of the 2-phase approach is evident.

In many situations vegetation surveys are purely based on field assessments. The data recorded for a single plot are extrapolated to the area represented by the plot; i.e., the surrounding area, and used to produce pixel maps. The grid density of 1-phase sampling designs has to be fairly large to give a realistic situation of the spatial distribution and frequency of species in the area under concern. Especially, the necessary sample size to detect sensitive (and meaningful) changes of spatial patters should not be underestimated. A stratification in aerial photographs provides a more realistic picture of the spatial distribution of vegetation types. If a 2-phase design is applied, cost efficiency for the estimation of current state and changes is better compared to 1-phase designs. Thus, whenever possible, the combination of field assessments and photo interpretation should be given preference.

ACKNOWLEDGMENTS

We would like to thank Dr. Bernhard Oester, WSL, for providing the results of the photo-interpretation. Dr. Risto Päivinen, European Forest Institute (EFI), and Dr. Peter Brassel, WSL, reviewed the manuscript and gave helpful comments.

REFERENCES

Broggi, M.F., Ed. *Inventar der Flachmoore von nationaler Bedeutung. Entwurf für die Vernehmlassung.* Bern, Bundesamt für Umwelt, Wald und Landschaft (BUWAL), 1990.

Cochran, W.G. *Sampling Techniques,* John Wiley & Sons, New York, 1977.

Gertner, G.Z. and M. Köhl. An assessment of some nonsampling errors in a national survey using an error budget. *For. Sci.,* 38(3) pp. 525–538, 1992.

Grünig, A., L. Vetterli, and O. Wildi. *Die Hoch- und Uebergangsmoore der Schweiz—eine Inventarauswertung.* Bericht Eidgenöss. Forsch. anst. Wald Schnee Landsch. 281, 1986.

Grünig, A., Ed. *Mires and Man. Mire Conservation in a Densely Populated Country—The Swiss Experience.* Excursion guide and symposium proceedings of the 5th field symposium of the International Mire Conservation Group (IMCG) to Switzerland 1992. Birmensdorf, Swiss Federal Institute for Forest, Snow and Landscape Research, 1994.

Kleinn, C. *Der Fehler von Flächenschätzungen mit Punktrastern und linienförmigen Stichproben.* Universität Freiburg, Mitteilungen der Abteilung Forstliche Biometrie. Nr. 91-1, 1991.

Köhl, M. *Statistisches Design für das zweite Schweizerische Landesforstinventar: Ein Folgeinventurkonzept unter Verwendung von Luftbildern und terrestrischen Aufnahmen.* Mitteilungen der Eidgenössischen Forschungsanstalt für Wald, Schnee und Landschaft, Birmensdorf, Band 69, Heft 1, 1994.

Köhl, M. and C. Kleinn. Stichprobeninventuren zur Waldflächenschätzung. *Allgemeine Forst- und Jagdzeitung.* 165 (12), pp. 229–231, 1994.

Küttel, M., Moorschutz in der Schweiz—Stand und Ziele. *Telma,* 25, pp. 177–192, 1995.

Schmid, P. Die Weiterentwicklung der Leistungskontrolle in der Schweiz. *Wissenschaftliche Zeitschrift der Technischen Universität Dresden.* Jg. 16/2, 1967.

Stott, C.B. Permanent Growth and Monitoring Plots in Half the Time. *J. For.* 37, pp. 669–673, 1947.

Chapter 5

The Effect of Uncertain Locations on Disease Cluster Statistics

Geoffrey M. Jacquez and Lance A. Waller

INTRODUCTION

Public health agencies use cluster statistics to assess whether an alleged aggregation of disease cases is statistically unusual. The health agency may choose to stop a cluster investigation when the alleged cluster appears to be due to chance, otherwise a more detailed and expensive investigation may be warranted (Centers for Disease Control, 1990). Health researchers typically work with two kinds of data: rates and points. Rates are defined over a spatial support (e.g., county) and time interval (e.g., year), and describe the number of health events (e.g., childhood leukemia cases) divided by the at-risk population (e.g., number of children in that county in that year). This chapter concerns itself with point data, which describe the locations and times of occurrence of individual health events. An example is the place of residence and time of diagnosis of a childhood leukemia patient. Whether they use rate- or point-data, disease cluster statistics seek to determine whether an excess of cases exists in space, in time, or in both space and time. When an excess exists it may be due to chance, but it might also be attributable to a common cause, such as exposure to an environmental carcinogen or to an infectious agent.

Point-based statistics assume exact geographic locations and are typically employed when health events are rare and the number of observations is small. They are coming into increasing use as spatially referenced health data become commonly available (Openshaw, 1991), and models of spatial point processes (Diggle, 1993; Lawson and Waller, 1996), become increasingly sophisticated. However, the assumption of exact locations upon which point-based statistics are founded is often invalid because our knowledge of health event location is uncertain. Does the violation of this assumption make a difference?

Sources of Location Uncertainty

It is well recognized that all GIS data are uncertain, and that uncertainty exists both in attributes and coordinates (see Heuvelink, 1998, and references therein). Here we are concerned with uncertainty in the coordinates, which we refer to as "location uncertainty." This is also called "positional uncertainty" in the literature, a usage that has gained currency even though it is less specific ("position" has 13 definitions in *Webster's New World Dictionary,* while "location" has 4) and unnecessarily adds the suffix "al" to create an adjective. Throughout this chapter we use "location uncertainty" to refer to uncertainty in a location's coordinates.

Most of the literature on uncertainty deals with uncertain attributes (Goodchild and Gopal, 1989; Heuvelink et al., 1989) and the propagation of attribute uncertainty through map operations (Haining and Arbia, 1993). In general, techniques for dealing with attribute uncertainty in statistical analyses are well developed (e.g., Viertl, 1996). In contrast, methods for assessing the amount of location uncertainty and its impact on outcomes such as spatial statistics have received little attention. Some research has dealt with location uncertainty and the calculation of lengths and areas (Keefer et al., 1991). Altman (1994) presents a fuzzy theoretic approach for representing location uncertainty, and Jacquez (1996a) used fuzzy set theory to develop disease cluster tests for imprecise locations. Using a probabilistic approach, Kiiveri (1997) presents a technique for assessing uncertainty in the locations of points and lines. Such probabilistic approaches can result in many potential realizations of the spatial data surface (see Goodchild et al., 1992; Ehlschlaeger and Shortridge, 1996, and references therein). This has motivated research on techniques for visualizing uncertainty, including animation (Ehlschlaeger et al., 1997).

To date and to our knowledge little, if any, research has dealt with the issue of propagating location uncertainty through spatial statistical analyses, and the consequences of location uncertainty on statistical inference. The research described in this chapter presents a technique for evaluating the impact of uncertain locations on disease cluster statistics. While the application is specific to the health sciences, the methodology is general and applies equally well to all fields of environmental science. While the remainder of our discourse focuses on health applications, the reader should be able extend the methods and results to his or her own scientific field without undue effort.

What are the sources of location uncertainty in health data? Point-based statistics assume data of the form (x, y, t) where x, y is the geographic coordinate (such as place of residence) and t is the date of onset, diagnosis, or death. More often than not this location model is inappropriate because locations of the health events are uncertain. This uncertainty has several sources. It emerges when centers of areas such as zip-code zones or census tracts are used instead of exact place of residence. Uncertainty also arises when data are gridded (as, for example, in raster-based GIS) and the coordinates of the nearest grid node are used instead of exact locations. Conceptually, locations are almost always uncertain because humans are mobile rather than sessile and because health events and their causative exposures may occur anywhere within a person's activity space. Tobler et al. (1995) observed that in modern society a person's daily activity space is approximately 15 km and varies widely. Exact locations do not represent this location uncertainty.

There are many examples of uncertain locations in the literature. Cuzick and Edwards (1990) used the centers of postal code zones to represent place of residence in their study of childhood leukemia in North Humberside. Waller et al. (1995) mapped locations of cases of childhood leukemia in Sweden at parish centroids, and evaluated proximity to nuclear power plants using several methods. Location uncertainty arises in these two examples because area centroids are used instead of exact locations. In a study notable for its spatial resolution, Lawson and Williams (1994) used place of residence to assess a possible cluster of deaths from respiratory cancer near a smelter in Armadale, Scotland. Indoor air quality is known to differ substantially from that outdoors, and an unknown proportion of each individual's exposure to smelter fumes presumably occurred outside of the home. In this example, location uncertainty arises because place of residence is used to represent exposures that occurred throughout each person's activity space.

We have not evaluated whether or not uncertain locations had an effect, if any, on these studies. Rather, our objective in citing these examples is to give readers a feeling for the ubiquity of the location uncertainty problem. In practice, geographic locations are, to a greater or lesser extent, inherently uncertain because they are proxy measures of exposure. If we had more information we could explore dose-response relationships using controlled studies. When such detailed knowledge is lacking we instead use spatial relationships among cases, and their proximity to putative sources of hazard, as an uncertain measure of exposure. However, almost all point-based cluster statistics assume the spatial coordinates of health events are precise. What is the impact, if any, of using statistics that assume exact locations when the locations are actually uncertain?

METHODS

Perhaps the most frequently used paradigm for dealing with uncertain locations is the centroid model. This model assigns all cases occurring within an area to the location of that area's centroid. It was used explicitly in the studies in Humberside and Sweden and is implied in the Armadale study. It obtains whenever data are gridded and is common in environmental epidemiology where insufficient knowledge of exposures and confidentiality concerns preclude exact knowledge of health event locations. Using a simulated AIDS epidemic in Michigan, we explored the performance of three disease cluster statistics at three levels of spatial resolution. We compared the test statistics, null distributions, P-values, and statistical power under the centroid model to those obtained when actual locations were used. This comparison allowed us to directly evaluate the impact of location uncertainty on the performance of disease cluster tests.

Cluster Statistics

Three statistics for space-time clustering were compared, Mantel (1967), Knox (1964), and k-NN (Jacquez, 1996a). These commonly used tests were chosen because they are based on different spatial proximity measures; geographic distance, adjacencies, and nearest neighbor relationships, respectively (Jacquez, 1996b). Using tests based on different proximity measures allows us to generalize our results to a wider range of cluster statistics. All cluster statistics were calculated using the Stat! software (Jacquez, 1994). These tests are now described.

Mantel's test (1967) is based on regression of the space distance between pairs of cases on the time distance between case pairs. Mantel reasoned that nearby cases may tend to occur at about the same time under a contagious disease process, or after an episodic release of a carcinogen from a point source. Assume x_i, y_i is the place of residence of the ith case, and t_i is the time of diagnosis. Define s_{ij} to be the Euclidean distance between the locations of cases i and j, and let w_{ij} be the waiting time (time distance) between case i and case j. Mantel's statistic is then

$$Z = \sum_{i=1}^{n}\sum_{j=1}^{n} s_{ij}w_{ij} \tag{1}$$

Here n is the number of cases (space-time locations). This statistic is sensitive to a relationship between the space and time distances such that large space distances are associated with large time distances, medium space distances with medium time distances, and so on. Under contagion, case locations that are far apart tend to be time independent, and Mantel suggested a reciprocal transformation which is the inverse of the space and time distances. This discounts the large space and time distances, and we used the reciprocal transformation suggested by Mantel. The reference distribution of the test statistic under the null hypothesis may be generated using the normal approximation of Mantel, or by a randomization that repeatedly "sprinkles" the t_i over the case locations, and recalculates Z each time. P-values are then determined by comparing the statistic for the not-randomized data to this null distribution. The null hypothesis for this test is that the space and time distances between case pairs are independent.

Knox (1964) used critical space and time distances to classify pairs of cases as either near or far in space and time. For example, two cases are considered near to one another when the distance between them is less than the space critical distance, and two cases are deemed near in time when their waiting time is less than the critical time distance. This recodes the data to be either "0" or "1," as follows. Define a time adjacency between cases i and j as $a^t_{ij}=1$, if $w_{ij}<\tau$, and 0 otherwise. Similarly, define a spatial adjacency between cases i and j as $a^s_{ij}=1$, if $s_{ij}<\delta$, and 0 otherwise. Here τ is the critical time distance (waiting time), and δ is the critical space distance.

Ideally, the critical distances are chosen to reflect the researchers' hypothesis regarding the epidemiological process at work. For example, when working with infectious diseases the critical time distance reflects the latency between exposure and the manifestation of symptoms, and the critical space distance to correspond to migration or movement during this latency period. The test statistic is

$$X = \sum_{i=1}^{n} \sum_{j=1}^{n} a^s_{ij} a^t_{ij} \tag{2}$$

Knox's X is sensitive to space-time clustering such that cases near in space are also near in time. The null hypothesis is that the space and time adjacencies are independent. The significance of the test statistic is evaluated as a chi-square or by randomization. In this study we selected critical values to maximize the Knox test statistic, thereby giving the test the "best chance" to maximize its statistical power. In field situations researchers don't have this luxury, and the results reported later for the Knox test are an upper bound on its performance.

k-NN test. Jacquez (1996a) recognized that selection of the constant for the reciprocal transformation in the Mantel test and of critical space and time distances in the Knox test is subjective. Further, he pointed out that population density varies, and a fixed critical spatial distance is therefore inappropriate. He proposed the k-NN statistic, which is the intersection of the space and time nearest neighbor relationships between cases pairs. Define k to be the number of nearest neighbors being considered. For example, the person nearest to you in a crowd is your first nearest neighbor, the next closest person is your second nearest neighbor and so on. Define $n^s_{ij}(k)$ to be 1 if case j is a k nearest neighbor in space of case i, and 0 otherwise. Further, define $n^t_{ij}(k)$ to be 1 if case j is a k nearest neighbor in time of case i, and 0 otherwise. The test statistic is

$$J(k) = \sum_{i=1}^{n} \sum_{j=1}^{n} n_{ij}^{s}(k) n_{ij}^{t}(k) \tag{3}$$

This is the count of the number of case pairs that are k nearest neighbors in both space and time. The statistic $J(k)$ is not independent of $J(k-1)$ since $J(k-1)$ case pairs are included in $J(k)$. This dependency is circumvented by the $\Delta J(k)$ statistic, which is

$$\Delta J(k) = J(k) - J(k-1); \; k = 2, \ldots \max(k) \tag{4}$$

The k-nearest neighbor statistics are sensitive to space-time clustering such that cases that are k nearest neighbors in space are also k nearest neighbors in time. Their significance is evaluated using randomization by permuting the t_i over the spatial locations.

Geography

Four levels of geographic uncertainty were evaluated: none, so that actual locations were used; county; bicounty; and region (Figure 5.1). Places of residence of the at-risk population are shown in Figure 5.1a by the "+" symbol and were sampled based on population density distributions of the global demography project (Tobler et al., 1995). Centroids and areas are shown for counties (Figure 5.1b), bicounties (Figure 5.1c), and regions (Figure 5.1d). Area centroids are shown as squares.

HIV Simulation

A simulated epidemic of AIDS in Michigan was used to generate space-time distributions of cases, and is based on the model of Jacquez et al. (1994). This HIV model was chosen because it describes the dynamics of the epidemic and because the transmission parameters have been estimated. The model was simplified for this research and, while the simplified version cannot be said to represent HIV in human populations, it is a contagious process and should result in space-time clusters. If the cluster statistics perform poorly for this model we have little reason to expect them to perform well for diseases such as environmentally caused cancer, for which space-time clustering should be substantially weaker.

The modeled disease has four stages: healthy, primary infection, seropositive, and deceased. The vector of transmission probabilities was (0.0, 0.3, 0.001, 0.0). Each element of this vector describes the probability, for a healthy individual, of infection per contact with an individual in disease stage i. The underlying population consists of 500 individuals with a heterogeneous geographic distribution (Figure 5.1a). Two of these 500 were selected as seeds to start the epidemic. At each time step, each healthy individual contacts one of any of four nearest neighbors, and a transmission event can occur when the contacted individual is infectious. In these instances the disease stage of the infected individual is determined and the corresponding transmission probability is used to determine whether infection occurs. For example, if the contacted individual is in primary infection, the corresponding value from the vector of transmission probabilities is 0.3. A uniform random number between 0 and 1.0 is then generated and an infection event (the healthy individual enters primary infection) occurs when this number is less than or equal to 0.3. Transition from

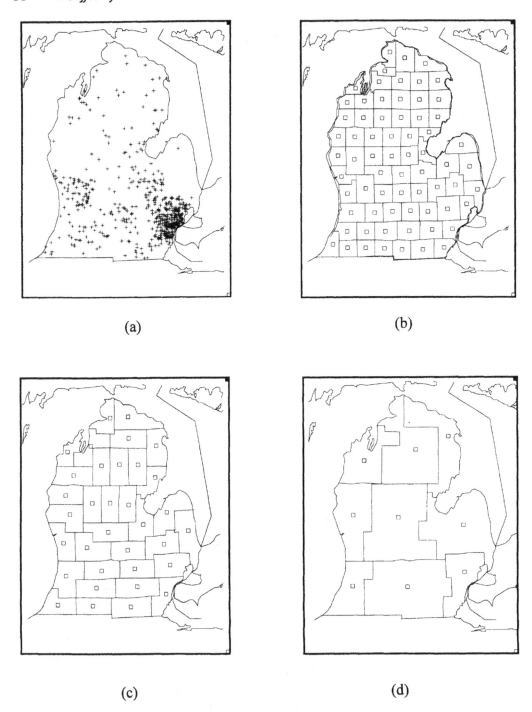

(a)   (b)

(c)   (d)

Figure 5.1. Boundaries corresponding to four levels of geographic uncertainty.

healthy to primary infection does not occur in the absence of a transmission event. Probabilities describe transition from one disease stage to another. In each round there is a 5%

probability that individuals in the primary infection stage will become seropositive, while, on average, 1% of seropositives enter the terminal stage.

The simulation is a three-step process; first transmission events (which cause healthy individuals to transit to primary infection) are evaluated. This involves determining each healthy individual's four nearest neighbors, and selecting which one of these four to have contact with. If the contacted individual is infected, whether a transmission event takes place is determined based on the vector of transmission probabilities as described above. Next, stage transitions are determined using transition probabilities. Finally, after each iteration is complete, the space-time location of each new seropositive is recorded and the simulation halts when 50 individuals become seropositive. The seropositive stage thus was used for the reporting of disease clusters. The model is stochastic and uses a uniform random number generator to evaluate the discrete transitions for each individual in the population. It is realistic because seropositives are monitored, and because stochasticity is introduced through probabilistic processes of contact, transmission, and state-to-state transition. As with all models, it is a simplification of reality designed to capture some of the variation and pattern that characterizes the real world.

Simulation Protocol

The HIV simulation model was run 252 times using the actual locations shown in Figure 5.1a, and new locations of the two initial infectives were chosen for each run. Two hundred fifty-two runs were chosen because this proved more than adequate for evaluating the level of association between actual and centroid test statistics. As described later, P-values for the actual and centroid test statistics are essentially independent, based both on visual inspection and Pearson product moment correlations (Figure 5.2). Each of the points in this figure represents a simulation run which is a distinct realization of the AIDS epidemic. We call each run an "AIDS simulation." Each AIDS simulation produced a list of coordinate pairs describing the actual locations of 50 seropositives. Centroid locations for the county, bicounty, and regional aggregation levels were assigned as appropriate (Figure 5.1). For example, at the county level a case occurring in county i was assigned the coordinates of that county's centroid. The actual test statistic (Γ_A) and reference (null) distribution (g_A) were calculated using the actual locations, and usually are not observable in field situations when centroids are used. The centroid test statistic (Γ_C) and its null distribution (g_C) were then calculated using the centroid locations, and correspond to statistics calculated in studies where centroid locations are used. The g_A and g_C distributions used 249 randomization runs to generate reference distributions expected under the null hypothesis of no association between a case's location and the time at which the case became seropositive. This was accomplished by reallocating the times the cases became seropositive at random across the case locations. These statistics and distributions were recorded for each AIDS simulation and for each cluster statistic.

In all, 252 AIDS simulations were run, and 249 Monte Carlo randomizations were used to generate each statistical null distribution. Thus 188,244 Monte Carlo randomizations were needed to generate the g_A distributions (252 AIDS simulation × 3 cluster tests × 249 randomizations) and 564,732 Monte Carlo randomizations were needed to generate the g_C distributions (252 AIDS simulation × 3 cluster tests × 3 uncertainty levels × 249 randomizations). The term "uncertainty level" refers to the level of aggregation (county, bicounty,

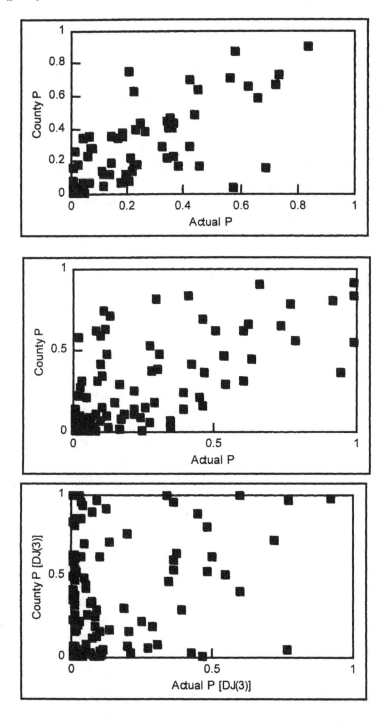

Figure 5.2. Results at the county level, centroid P-values vs. actual P-values for Mantel (top) Knox (middle), and k-NN (bottom) tests. For the k-NN test the test statistic was maximized when the number of nearest neighbors considered was three (k=3). For this reason P-values for this statistic at k=3 [DJ(3)] are shown. Scatter on the plots for different values of k was greater.

region) used to assign centroid locations. A grand total of 752,976 Monte Carlo randomizations were conducted using spatial uncertainty modeling software being developed by BioMedware.

RESULTS

Inquiry focused on four questions. How does uncertainty regarding the spatial locations of health events affect (1) the test statistic, (2) its null distribution, (3) P-values, and (4) statistical power of point-based cluster statistics?

Figure 5.2 shows the results at the county level. For Mantel's test, test statistics calculated from the centroid locations were consistently lower than those calculated using actual locations, and the upper 95% critical value of the g_C reference distribution was not highly correlated with the 95% critical value of the $g_A>$ reference distribution calculated from the actual locations. As a result, P-values using the centroid locations were essentially independent of P-values based on the actual locations (Figure 5.2a). This pattern of results holds for both the Knox test (Figure 5.2b) and the k-NN test (Figure 5.2c). Results, at the bicounty and regional aggregation levels are as bad or worse and demonstrate, for the HIV model and at the spatial scales considered, that P-values obtained under the centroid location model differ substantially from P-values based on actual locations.

For the three cluster statistics considered there is a substantial decrease in statistical power at the county, bicounty, and regional aggregation levels. Statistical power was quantified as the proportion of simulation runs for which the P-value was less than or equal to 0.05, meaning the correct finding of "significant space-time clustering" was found. Using actual locations the statistical power was 0.592, 0.480, and 0.792 for the Mantel, Knox, and k-NN tests, respectively. Using county centroids the drop in statistical power ranged from 3.67% (Knox) to 36.62% (k-NN). At the bicounty level the drop ranged from 6.58% (Knox) to 44.1% (k-NN). At the regional level the drop in statistical power ranged from 22.60% (Knox) to 59.83% (k-NN).

These results illustrate that P-values from centroid data bear little resemblance to P-values based on actual locations. This causes a loss in statistical power when using centroid locations, and an increase in type II error (false negatives).

CONCLUSION

The change in P-values when centroids are used is of itself not surprising, and is consistent with the results of Waller and Turnbull (1993), who reported changes in P-values at the census tract and block group aggregation levels for focused cluster tests. What is disturbing is the decrease in statistical power. As mentioned in the introduction, cluster statistics are one of several criteria used by public health agencies to determine the future course of a cluster investigation. In such situations, P-values from disease cluster tests are used to quantify how unusual an observed aggregation of cases is. A very small P-value suggests the aggregation was unlikely to have been a chance event, and additional effort may be needed to identify environmental exposures that might explain the cluster. While generalization from a single simulation study is difficult, our results suggest that P-values based on uncertain locations are unreliable and should not be used as a quantitative basis for determining the future course of cluster investigations.

Several authors have commented on the apparent lack of statistical power in cluster investigations, and some reviewers of this chapter were surprised that the power of the tests was relatively low, even when the actual locations were used. A test's power to detect departures from the null hypothesis depends on many things, including sample size, and the strength of the departure from the null model. In this chapter we endeavored to use a realistic space-time disease model, actual population density distributions, and the relatively small sample sizes typical of cluster investigations. If one accepts these conditions as representative of reality, our results are discouraging from a public health perspective because the proportion of false negatives is substantial. This study strongly suggests that actual clusters are being missed because of the loss of statistical power due to location uncertainty.

This conclusion is interesting to consider within the context of the "Texas sharpshooter" syndrome, which refers to cluster preselection bias. The Texas sharpshooter aims at a barn, shoots, and then draws a bull's-eye around the bullet hole. Being environmentally aware and concerned about health, the public is often viewed as a Texas sharpshooter when it brings forward putative disease clusters. Most statisticians and health professionals accept that such preselection bias exists, and further believe that it should lead to an increase in false positives. This increase in false positives is expected because the public is constantly surveying for apparent health clusters, and raises an alarm even when the apparent cluster is a chance aggregation. On the other hand, anecdotal evidence from "cluster busters" strongly suggests that the detection rate (number of cluster investigations with a P-value less than 0.05, divided by the total number of cluster investigations) is very low, on the order of fewer than 1 in 100. Notice if the clusters are due to chance alone we expect positive results at the alpha level, or 5 in 100. And if preselection bias exists, it should be even higher. But the detection rate is too low, not too high! We call this the Texas sharpshooter paradox: preselection bias exists, yet the statistical evidence is to the contrary. It is as if the Texas sharpshooter is firing blanks, and drawing circles anyway. A possible explanation is the all-too-frequent use of centroid locations in disease cluster investigations. Centroids are hyper-dispersed and violate the Poisson assumption on which most point-based statistics are founded (Diggle, 1993). This makes it more difficult to detect clustering since it must overcome the regular pattern imposed by hyperdispersion.

Even when place of residence is known to the level of the street address, geographic location is often uncertain because health events occur throughout a person's daily activity space, and because exposures (e.g., transmission events for infectious diseases, exposure to radiation and carcinogens for cancer, etc.) underlying health events occur outside as well as inside the home. This indicates a substantial need for ways of accounting for uncertain locations in general, and for point-based statistics in particular.

There are broader implications because the cluster tests considered are similar to statistics used in other fields. For example, nearest neighbor statistics are used extensively in ecology, and Mantel's test can be expressed as the Pearson product moment correlation which is used extensively in the environmental sciences. All GISs employ data models that, depending on the GIS operation, may abstract locations to grid nodes or polygon centers, thereby introducing location uncertainty. This study therefore may have implications for statistical analyses of data managed by Geographic Information Systems.

In conclusion, the methods and software developed in this research allow one to assess, given a specific geography and hypothesized space-time process, the impact of uncertain locations on test statistics, reference distributions, P-values, and statistical power. Generalization from the single simulated disease process and geography considered is problematic. Will the pattern of results hold at finer geographic resolutions? Because of their use in small area studies, it seems particularly important to assess the impact of uncertain locations for census tract and zip-code zone geography. Now that a substantial effect at the county level has been demonstrated, the methods developed in this preliminary study can be used to address this question and others like it.

ACKNOWLEDGMENTS

This research was funded by Small Business Innovation Research grants R43 CA65366-01A1 and R43 CA65366-02 from the National Cancer Institute. Its contents are solely the responsibility of the authors and do not necessarily represent the official views of the NCI.

REFERENCES

Altman, D. Fuzzy set theoretic approaches for handling imprecision in spatial analysis. *Int. J. Geogr. Inf. Syst.,* 8, pp. 270–289, 1994.

Centers for Disease Control. Guidelines for investigating clusters of health events. *Morbidity Mortality Wkly. Rep.,* 39, pp. 1–23, 1990.

Cuzick, J. and R. Edwards. Spatial clustering for inhomogeneous populations. *J. R. Stat. Soc.,* 52, pp. 73–104, 1990.

Diggle, P.J. Point process modelling in environmental epidemiology. *Statistics for the Environment.* V. Barnett and K.F. Turkman, Eds., John Wiley & Sons Ltd, New York, 1993, pp. 89–110.

Ehlschlaeger, C. and A. Shortridge. *Modeling Elevation Uncertainty in Geographical Analysis.* Spatial Data Handling '96, Delft, The Netherlands, 1996.

Ehlschlaeger, C.R., et al. Visualizing spatial data uncertainty using animation. *Comput. Geosci.,* 23(4), pp. 387–395, 1997.

Goodchild, M. and S. Gopal. *Accuracy of Spatial Data Bases.* Taylor and Francis, New York, 1989.

Goodchild, M.F., et al. Development and test of an error model for categorical data. *Int. J. Geogr. Inf. Syst.,* 6(2), pp. 87–104, 1992.

Haining, R.P. and G. Arbia. Error propagation through map operations. *Technometrics,* 35, pp. 293–305, 1993.

Heuvelink, G.B.M. *Error Propagation in Environmental Modelling with GIS.* Taylor and Francis, London, 1998.

Heuvelink, G.B.M., et al. Propagation of errors in spatial modeling with GIS. *Int. J. Geogr. Inf. Syst.,* 7, pp. 231–246, 1989.

Jacquez, G.M. *Stat! Statistical Software for the Clustering of Health Events.* BioMedware, Ann Arbor, MI, 1994.

Jacquez, G.M. Disease cluster statistics for imprecise space-time locations. *Stat. Med.,* 15, pp. 873–885, 1996a.

Jacquez, G.M. A *k*-nearest neighbor test for space-time interaction. *Stat. Med.,* 15, pp. 1935–1949, 1996b.

Jacquez, J.A., et al. Role of the primary infection in epidemics of HIV infection in gay cohorts. *J. Acquired Immune Deficiency Syndr.,* 7, pp. 1169–1184, 1994.

Keefer, B.J., et al. Modeling and evaluating the effects of stream mode digitizing errors on map variables. *Photogrammetric Eng. Remote Sensing.,* 57, pp. 957–963, 1991.

Kiiveri, H.T. Assessing, representing and transmitting positional uncertainty in maps. *Int. J. Geogr. Inf. Sci.,* 11(1), pp. 33–52, 1997.

Knox, G. The detection of space-time interactions. *Appl. Stat.,* 13, pp. 25–29, 1964.

Lawson, A.B. and L.A. Waller. A review of point pattern methods for spatial modelling of events around sources of pollution. *Environmetrics,* 7, pp. 471–487, 1996.

Lawson, A.B. and F.L.R. Williams. Armadale: A case-study in environmental epidemiology. *J. R. Stat. Soc.,* 157, pp. 285–298, 1994.

Mantel, N. The detection of disease clustering and a generalized regression approach. *Cancer Res.,* 27, pp. 201–218, 1967.

Openshaw, S. A new approach to the detection and validation of cancer clusters: A review of opportunities, progress and problems. *Statistics in Medicine,* F. Dunstan and J. Pickles, Eds., Clarendon Press, Oxford, 1991, pp. 49–63.

Tobler, W., et al. *The Global Demography Project.* Santa Barbara, National Center for Geographic Information and Analysis, 1995.

Viertl, R. *Statistical Methods for Non-Precise Data.* CRC Press, Boca Raton, FL, 1996.

Waller, L.A. and B.W. Turnbull. The effects of scale on tests of disease clustering. *Stat. Med.,* 12, pp. 1869–1884, 1993.

Waller, L.A., et al. Detection and assessment of clusters of disease: An application to nuclear power plant facilities and childhood leukemia in Sweden. *Stat. Med.,* 14, pp. 3–16, 1995.

Chapter 6

Double Sampling for Area Estimation and Map Accuracy Assessment

Francisco Javier Gallego

INTRODUCTION

Area frame sampling is a major tool for area estimation. Many approaches can be considered: sampling points, transects, or aerial units (Webster and Oliver, 1990). The territory is often stratified to optimize sampling; this can be achieved by photo-interpretation of satellite images (Cotter and Tomczac, 1994; Avenier et al., 1992). When the studied region is very large, analyzing a complete coverage of high resolution satellite images may be too expensive and double sampling can provide an efficient alternative to stratified sampling.

In order to avoid confusion we adhere to the definition of two-stage and two-phase (or double) sampling as given by Marriot (1990):

- Two-stage sampling: clusters of units are sampled in the first stage. In the second stage elementary units are sampled inside the selected clusters.
- Two-phase (or double) sampling: a large sample of units are selected in the first phase; auxiliary information is obtained for the first phase sample as a basis to draw a stratified or unequal probability sample in the second phase.

Both techniques can be combined in different ways; for example, double sampling can be applied in the first or the second stage of a two-stage sampling.

We illustrate the use of double sampling, alone or combined with two-stage sampling, with examples of projects in which the Commission of the European Community (CEC) is involved, considering three types of linked problems:

- area estimation (land use or land cover)
- area change estimation between two reference dates (for example, deforestation)
- map accuracy assessment; i.e., estimation of disagreement between a land cover map and reference data on a sample.

SAMPLING SITES OF 40 KM × 40 KM FOR CROP AREA CHANGE ESTIMATION

Rapid crop area estimates are being produced in the MARS Project (Monitoring Agriculture with Remote Sensing, Meyer-Roux and King, 1992; Meyer-Roux and Vossen, 1994)

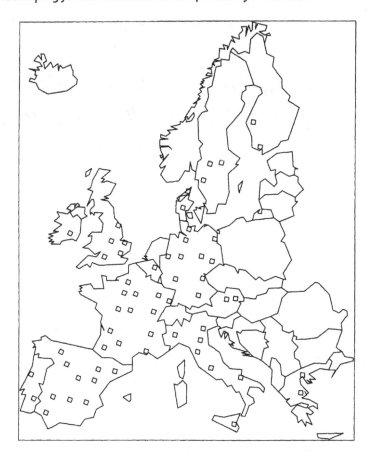

Figure 6.1. Sample of 60 sites in the European Union.

for the European Union (EU) by multitemporal analysis of high resolution satellite images on a sample of 60 sites of 40 km × 40 km (Figure 6.1). In each site 16 segments of 1400 m × 1400 m are ground surveyed. More details on the complete sampling design can be found in Carfagna et Gallego (1997).

The unit size and sample distribution was subjectively chosen in 1988 because:

- No data were available at that moment to assess cost-efficiency of the unit size. A 40 km × 40 km square seemed convenient because it comfortably fits inside the high resolution satellite images that were targeted for the project: SPOT-XS (60 km × 60 km) and Landsat-TM (180 km × 180 km).
- The original aim was "crop status assessment." Difficulties arose later to extrapolate results to the EU level when area change estimates were considered feasible.

Sampling Sites in Central and Eastern European Countries

In 1994, a sample was requested to extend the "rapid crop area estimates" to six countries in Central Europe (Poland, Czech Republic, Slovakia, Hungary, Romania, and Bul-

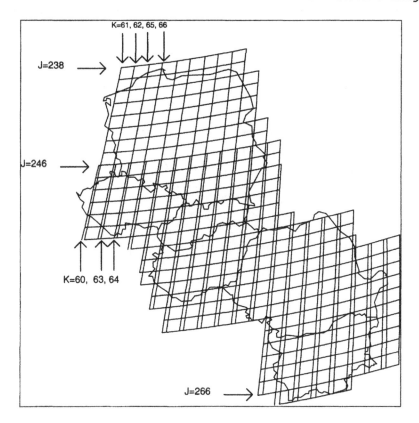

Figure 6.2. SPOT reference grid in Central Europe.

garia). A two-phase (double) sampling has been followed to select 40 sites to avoid the drawbacks experienced in the earlier EU project.

First Sampling Phase

A systematic sample is selected by blocks on the K-J indexes of the SPOT reference grid (Figure 6.2). K is linked with the satellite orbit, and J refers to the latitude. The area was divided into blocks containing 8 Spot KJ (in the south of Europe) or 4 Spot K-J, north of 51°, where one track out of two disappears in the SPOT grid. One site is selected in each block. Sites on the border are selected if more than 50% is inside the studied region. A first phase sample (or presample) of $n' = 58$ sites was selected (Figure 6.3). The sampling probability π'_s in the first phase is higher in the north of the area because satellite orbits are closer to each other:

$$\pi'_s \propto \frac{1}{\eta_{lat(s)}}$$

where $\eta_{lat(s)}$ is the distance between contiguous orbits.

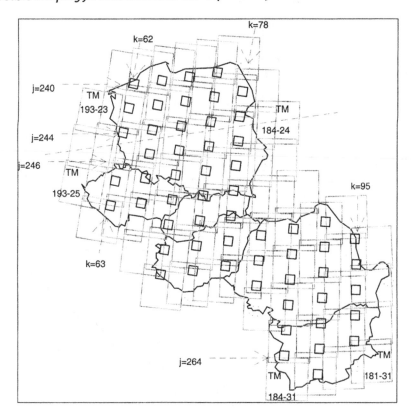

Figure 6.3. Systematic sample of sites based on the SPOT reference grid.

Second Sampling Phase

Spot XS digital quick-looks or Landsat-TM photographic printouts of the 58 sites in the presample were analyzed, giving a rough estimation A_s of the percentage of agricultural land. A_s is always a multiple of 10%. A second phase, or final sample of 40 sites (Figure 6.4) was subsampled without replacement with a probability $\pi_s'' \propto A_s \eta_{lat(s)}$ following a systematic πps selection method (Särndal et al., 1992, p. 96). The final sampling probability associated with each site is $\pi_s = \pi_s' \times \pi_s'' \propto A_s$.

Estimates of the area Z for land cover c are computed with a Horvitz-Thompson or π estimator (Särndal et al., 1992; Thompson, 1992) using the proportion y_s of land cover c in site s rather than the area z_s. This means that incomplete sites in the borders of the region (> 50% inside) weigh as much as complete sites and compensate in some way the fact that pieces smaller than 50% have disappeared from the area frame. Additional analysis is still necessary for a better assessment of this compensation.

$$\hat{y}_\pi = K' \sum_s \frac{y_s}{\pi_s} = K \sum_s \frac{y_s}{A_s} \qquad \hat{Z} = D\,\hat{y}_\pi$$

where D is the total area of the region.

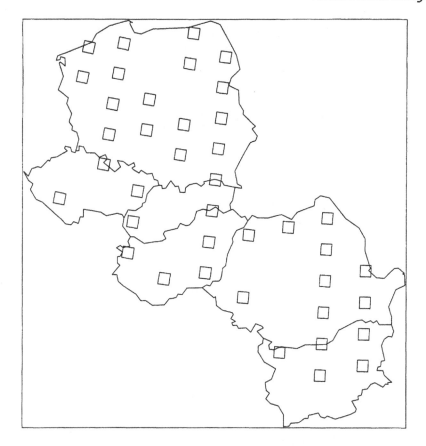

Figure 6.4. Sample of sites for image analysis.

The proportionality factor K can be computed considering the class "any land cover" that occupies 100% of the territory:

$$y_s = 1 \; \forall s \quad \Rightarrow \quad K = \sum_s \frac{1}{A_s}$$

The variance of the π-estimator often presents stability problems (Särndal et al., 1992). Difficulties are smaller if the probability π_s that the element s is in the sample has few different possible values, two-phase sampling can be viewed as a stratified sample in which the size of each stratum is not exactly known, but estimated from the first phase sample (Cochran, 1977): The set of elements sampled with a given probability represents a stratum that we call $\Omega_h = \{s/\pi_s = \pi_h\}$. We ignore the exact size of the stratum Ω_h but we can estimate it from the proportion of elements in the presample that are subsampled with probability π_h. An unbiased estimate for the mean is:

$$\bar{y}_{st} = \sum_h w_h \bar{y}_h$$

where w_h is the proportion of stratum Ω_h estimated from the presample: $w_h = n'_h/n'$

The variance can be estimated with the standard formula for two-phase random sampling (Cochran, 1977) :

$$v(\bar{y}_{st}) = \frac{n'(N-1)}{(n'-1)N} \sum_h w_h s_h^2 \left(\frac{1}{n'v_h} - \frac{1}{N}\right) + \frac{g'}{n'} \sum_h s_h^2 \left(\frac{w_h}{N} - \frac{1}{n'v_h}\right) + \frac{g'}{n'} \sum_h w_h (\bar{y}_h - \bar{y}_{st})^2$$

where

$$s_h^2 = \frac{1}{n_h - 1} \sum_{i \in \Omega_h} \left(y_{i-}\bar{y}_h\right)^2 \qquad n_h = \frac{n'_h}{n_h}$$

and

$$g'_h = \frac{N - n'}{N - 1}$$

The sampling technique presented here is closer to a systematic sampling, which generally gives a lower variance than random sampling (Das, 1950; Bellhouse, 1977, 1988; Dunn and Harrison, 1993); hence the formula for random sampling gives a conservative estimate (overestimates) of the variance.

SAMPLING SMALLER UNITS

Sampling units of 40 km × 40 km can be too large. With this size, only a small number of sites can be treated with a reasonable cost. Square units with a side length of 3 km to 10 km might allow a larger number of units to be managed at a similar cost. The applicability of this approach is linked with image distribution policy. Some European image vendors of Landsat TM and IRS-1C images are considering the delivery of small image windows with predefined coordinates. Smaller units would also give higher probability of cloud-free satellite images, and this might be a considerable advantage. Images with a 50% cloud cover, that are now seldom exploited, might become usable.

Correlograms computed on ground survey segments are useful to study the optimal size of sampling units. When we estimate interannual crop area change in the "rapid area estimates," correlograms usually have very low values, even at short distances. Figure 6.5 shows a smoothed correlogram for common wheat (the most important crop in western Europe) computed from 3,400 quarters of segments (700 m × 700 m). Using quarters of segments allows one to compute reliable correlograms at short distances (700 m and 700 m × $\sqrt{2}$), which are essential to study the optimal unit size.

The low values of correlograms at very short distances suggest that there is little redundancy between adjacent segments. Hence collapsing the 16 segments sampled in a site into a single macrosegment of 5,600 m × 5,600 m would not result in a major loss of information. Ongoing studies (Gallego, 1998) linking the correlogram and intracluster correlation (Cochran, 1977) support this conjecture.

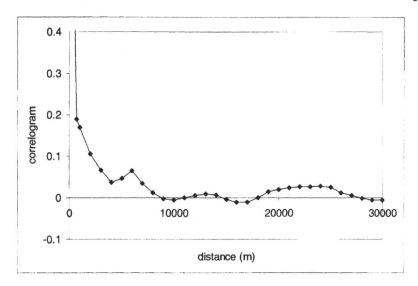

Figure 6.5. Correlogram for interannual wheat area change. Segments of 700 m.

Sampling Segments in Satellite Image Strips

In this two-stage approach the site (first stage sampling unit) is a strip along a satellite orbit track (Figure 6.6). Different sets of sites can be defined for different satellite sensors, such as Landsat-TM, SPOT-XS, or IRS-1C. Segments (second stage sampling units) are sampled inside the site. This strategy may become attractive to work with the very high resolution images that soon should be available. Several satellites foresee images with a resolution comparable to aerial photographs, but limited swath (6 or 8 km). Sampling strips compatible with the swath can be a good solution, although the fact that sites are linked to a particular satellite may be a serious drawback.

Segment sampling inside the strips (second stage) follows a two-phase procedure, while for the example of site sampling in Central and Eastern European countries the two-phase sampling was applied in the first stage. In the simulations presented below we have used systematic sampling for the first phase of this second stage, taking one segment every 10 km. In the second phase, a subsample is selected with a probability that depends on a proxy variable (an index of agricultural intensity, proportion of forest, or land cover variability).

COMPARING THE EFFICIENCY OF SAMPLING STRATEGIES

We do not yet have data to test the efficiency of the strip approach with real strips following satellite tracks, but we can simulate the scheme with data from ground survey segments. The set of data used for this test comes from 1992 and 1993 ground surveys on 8,023 square segments of 49 ha from the Spanish Ministry of Agriculture in an area of 270,000 km² (Castilla y León, Madrid, Castilla-la Mancha, and Andalucía). The set of 8,023 segments was obtained by systematic sampling with three replicates in blocks of 10 km × 10 km. The sample is described by MAPA (1990), Ambrosio et al. (1993), and

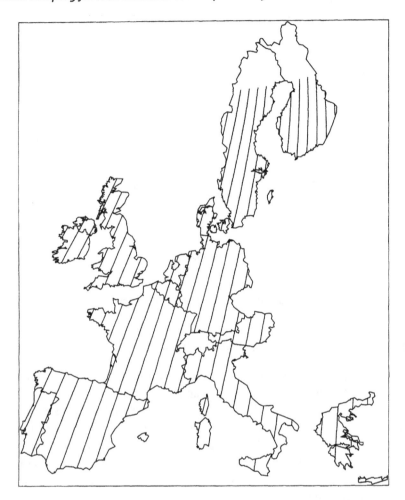

Figure 6.6. Strips on Landsat tracks.

Gallego (1995). We consider this set of 8,023 segments as population from which we select samples of 80 segments with different strategies. For each strategy, we have repeated the procedure 100 times in order to estimate variances. We compare relative efficiency values of area estimates in 1993. Relative efficiency is defined as

$$\text{relative efficiency} = \frac{\text{estimate variance with simple random sampling}}{\text{estimate variance with current strategy}}$$

Two strategies use an agricultural intensity index computed from CORINE Land Cover (CEC, 1993; Perdigão and Annoni, 1997), an environmental database obtained by photo-interpretation of high resolution satellite images, available in nearly all western and central European countries with common technical specifications.

To compute this agricultural intensity index, the CORINE Land Cover nomenclature was first simplified to 12 classes, a GIS layer was cut into squares corresponding to the

8,023 ground survey segments, and the area of each class measured for each segment. A linear regression between the total arable land observed in 1992 and the agricultural classes as mapped by CORINE suggested the following index:

$$\text{CORINE index} = \text{arable} + \frac{\text{vineyard} + \text{heterogeneous agriculture}}{4}$$

To estimate the variance of proportional probability subsampling using this index, 1,000 repetitions were necessary to reach stability.

Sampling Strategies Compared

- *Random:* simple random sampling without replacement. This is the reference to compute the relative efficiency of the other strategies.
- *Systematic uniform:* systematic sampling ordering population segments by geographic coordinates (latitude and then longitude). Uniform probability.
- *Proportional ground 92:* sample with a probability proportional to arable land observed in the ground survey in 1992.
- *Strips ground 92:* two-stage sampling: systematic sampling of strips 100 km apart from each other and subsample with probability proportional to 1992 arable land (Figure 6.7).
- *Strata ground 92:* four strata were defined on the basis of the 1992 arable land rate: <20%, 20–50%, 50–80%, and >80%. Sampling rates were proportional to 1, 2, 3, and 4, respectively.
- *Proportional CORINE:* sample with a probability proportional to the CORINE index described above.
- *Strata CORINE:* four strata were defined on the basis of the CORINE index.

Comments on Simulation Results

Some conclusions can be drawn by examining Table 6.1; in this we give the relative efficiency of the studied strategies compared with simple random sampling for the main annual crops and fallow (understood as arable land noncultivated in the current year):

Systematic sampling is superior to random sampling. This confirms once again the generally superior performance of systematic sampling compared with random sampling, in spite of the difficulties in computing their variance from a single sample.

Sampling with probability proportional to an index of agricultural intensity is much more efficient than simple random or systematic sampling if the index is very accurate, but has major risks if the index is moderately inaccurate. The reason is the huge contribution to the variance of segments wrongly characterized as "nearly nonagricultural," that have a low π value (high extrapolation factor) and a high crop area value.

Sampling by strips is not as good as sampling unclustered segments (in both cases with probability proportional to an accurate index), but may be an efficient strategy when working with very high resolution images.

The efficiency gain with a two-phase sampling can be higher, if subsampling probabilities are cautiously selected, than the efficiency obtained in most European countries with a stratification obtained by an ad-hoc photo-interpretation (Taylor et al., 1997).

Table 6.1. Relative Efficiency Compared With Random Sampling

	Systematic Uniform	Strips Ground 92	Proportional Ground 92	Strata Ground 92	Proportional CORINE	Strata CORINE
Wheat 93	1.20	2.80	6.53	3.60	0.27	2.21
Barley 93	1.61	4.93	6.65	3.02	1.27	1.51
Cereals 93	1.42	5.80	7.07	4.34	0.60	2.18
Sunflower 93	1.26	1.69	3.53	2.16	0.67	1.74
Sugar beet 93	1.14	1.69	2.85	2.28	2.28	1.74
Fallow 93	1.05	2.11	3.96	1.30	0.52	1.18

Subsampling in a few strata moderates the effect of proportional probability susbsampling, and should be preferred unless the associated index (proxy variable) is very reliable.

USING AVAILABLE GROUND SURVEY DATA TO ASSESS CORINE LAND COVER

One of the topics discussed in the previous section is the use of CORINE Land Cover to improve crop area estimates from a ground survey on an area frame sample. The reciprocal approach is also possible: available data from an agricultural ground survey can be useful to assess the accuracy of a land cover map. We shall illustrate here the use of the survey by the Spanish Ministry of Agriculture to assess CORINE Land Cover.

Inaccuracy is usually measured as the disagreement between the data set under assessment and a reference data set that we assume to be more accurate; for example, a ground survey. In general, reference data are only available for a sample. In this case the ground survey data cannot be directly used for accuracy assessment: a disagreement between CORINE Land Cover and the agricultural ground survey does not necessarily mean an error, for several reasons:

- nomenclatures are not fully compatible,
- the reference date of the current version of CORINE Land Cover in Spain is 1986–87,
- CORINE is not supposed to map landscape units smaller than 25 ha,
- location disagreement may be due to the ground survey data.

Even with these limitations, data from agricultural ground surveys are available and handling them is cheap. Here we propose a way to exploit this data following a two-phase sampling approach outlined in the following steps:

1. Consider the available sample as first stage sample (presample).
2. A disagreement index is computed for each segment of the sample. In our case we have defined:

$$\delta = \sqrt{\Delta\text{Arab}^2 + \Delta\text{Perm}^2 + \Delta\text{Grass}^2 + \Delta\text{Forest}^2 + \Delta\text{Other}^2}$$

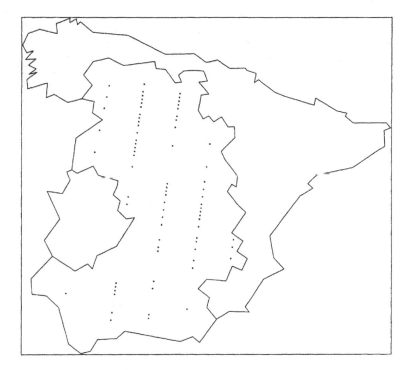

Figure 6.7. Simulated double sample on strips in Spain.

Where ΔArab=%Arable(CORINE)–%Arable(Ground Survey) and Δ*Perm,* Δ*Grass,* Δ*Forest,* Δ*Other* are defined in a similar way for permanent crops, grassland, forest, and other land cover.

3. Select a subsample of segments with a probability that depends on the disagreement index δ.
4. Collect ground information with a nomenclature compatible to CORINE Land Cover that leads to a proper confusion matrix.

For this exercise we considered a sample of 10,674 segments surveyed in 1993 on a region of 350,000 km^2. The maximum value of the disagreement is $\delta=100\sqrt{2}=141$. The segments were classified into four strata (Table 6.2) and a subsample was selected for a more careful analysis with higher rates for segment with a higher disagreement index, where problems are more likely to appear.

An appropriate assessment of the CORINE Land Cover 1986–87 database with synchronic ground visits to this sample is obviously not possible. The same sample can be recommended for the assessment of future updates of CORINE Land Cover, since it is probable that more errors occur in a future update in areas where photo-interpretation difficulties were apparent in 1987.

CONCLUSIONS

Two-phase sampling (double sampling) may be efficient for a variety of projects involving remote sensing applied to land cover area estimation or area change estimation. This is

Table 6.2. Stratification of a Large Sample of Segments in Spain According to a Disagreement Index between CORINE Land Cover and Ground Survey

Disagreement Index	Number of Segments	Subsampled
0–40	4314	43
40–80	2675	54
80–120	2141	64
>120	1544	62
Total	10674	223

especially true if no suitable stratification is available and/or if the area of study is too large and a complete coverage of high or medium resolution satellite images is not affordable.

When the accuracy of a land cover map is to be assessed, available information on a large sample, even if it is nonoptimal, may be used through double sampling to concentrate quality checks where errors are more likely to happen.

Double sampling with probability proportional to an auxiliary index can be very efficient if the ancillary information to compute the index has a high quality; otherwise, the effect of unequal probability sampling should be moderated by grouping index values into a few intervals, that define strata.

Sampling plans by long thin strips following satellite orbit tracks can be an efficient way of exploiting very high resolution satellite images with a narrow swath.

ACKNOWLEDGMENTS

We are grateful to the Spanish Ministry of Agriculture, in particular to Porfirio Sánchez, José María Fernández del Pozo, and María José Postigo, that has kindly provided data to test the efficiency of sampling strategies. Data provided by the CORINE Land Cover Task Force (European Environment Agency and European Commission, DG XI) were also essential for the results presented. Two anonymous reviewers made very useful comments to improve the chapter.

REFERENCES

Ambrosio, L., R. Alonso, and A. Villa. Estimación de superficies cultivadas por muestreo de áreas y teledetección. Precisión relativa. *Estadística Española*, 35, pp. 91–103, 1993.

Avenier, D., V. Perdigão, and J.M. Terres. *Méthodologie de stratification sur images satellitaires et utilisation d'un système d'information géographique; mise en place du plan d'échantillonnage.* Agriculture series. JRC Ispra, 1992.

Bellhouse, D.R. Systematic sampling, *Handbook of Statistics*, P.R. Krisnaiah and C.R. Rao, Eds. North-Holland, 1988, pp. 125–146.

Bellhouse, D.R. Some optimal designs for sampling in two dimensions, *Biometrika* 64, pp. 605–611, 1977.

Carfagna, E. and F.J. Gallego. Yield Estimates from Area Frame at European Level. *Seminar on Crop Yield Forecasting Methods. Villefranche sur Mer, 24–27 October 1994.* Office for Publications of the E.C. Luxembourg, 1997, pp. 197–202.

CEC. *CORINE Land Cover; Guide Technique,* Office for Publications of the European Communities. Luxembourg, 1993.

Cochran, W. *Sampling Techniques*. John Wiley & Sons, New York, 1977.

Cotter, J. and C. Tomczac. An image analysis system to develop area sampling frames for agricultural surveys. *Photogrammetric Eng. Remote Sensing*, 60(3), pp. 299–306, 1994.

Das, A.C. Two-dimensional systematic sampling and associated stratified and random sampling. *Sankhya*, 10, pp. 95–108, 1950.

Dunn, R. and A.R. Harrison. Two-dimensional systematic sampling of land use. *JRSS-C: Appl. Stat.* 42(4), pp. 585–601, 1993.

Gallego, F.J. *Sampling Frames of Square Segments*, Report EUR 16317, Office for Publications of the E.C. Luxembourg, 1995.

Gallego, F.J. *Geographic Sampling Strategies and Remote Sensing,* Report to Eurostat, 1998.

MAPA. El Marco de Areas como Instrumento de base para la Estadística de superficies de cultivo. *Boletín Mensual de Estadística Ministerio de Agricultura* n. 12, Madrid, 1990, pp. 85–106.

Marriot, F.H.C. *A Dictionary of Statistical Terms, 5th edition*. Longman, Harlow, 1990.

Meyer-Roux, J. and C. King. European Achievements in remote sensing: Agriculture and forestry, *Int. J. Remote Sensing*, 13(6–7), pp. 1329–1341, 1992.

Meyer-Roux, J. and P. Vossen. The First Phase of the MARS Project, 1988–1993: Overview, Methods and Results. *Conference on the MARS Project*. Office for Publications of the E.C., Luxembourg. pp. 33–81, 1994

Särndal, C.E., B. Swenson, and J. Wretman. *Model Assisted Survey Sampling*. Springer Verlag, 1992.

Perdigão, V. and A. Annoni. *Technical and Methodological Guide for Updating CORINE Land Cover Data Base,* Report EUR 17288, Office for Publications of the E.C. Luxembourg, 1997.

Taylor, J.C., C. Sannier, J. Delincé, and F.J. Gallego. *Regional Crop Inventories in Europe Assisted by Remote Sensing*. Report EUR 17319, Office for Publications of the E.C. Luxembourg, 1997.

Thompson, S.K. *Sampling,* John Wiley & Sons, New York, 1992.

Webster, R. and M.A. Oliver. *Statistical Methods in Soil and Land Resource Survey*, Oxford University Press, 1990.

Chapter 7

Accuracy Assessments and Areal Estimates Using Two-Phase Stratified Random Sampling, Cluster Plots, and the Multivariate Composite Estimator

Raymond L. Czaplewski

INTRODUCTION

Consider the following example of an accuracy assessment. Landsat data are used to build a thematic map of land cover for a multicounty region. The map classifier (e.g., a supervised classification algorithm) assigns each pixel into one category of land cover. The classification system includes 12 different types of forest and land cover: black spruce, balsam fir, white cedar, other softwoods, aspen, birch, other hardwoods, urban, wetland, water, pasture, and agriculture. The accuracy of the map must be known by a user of the map to conduct credible analyses.

In concept, each and every pixel can be classified with two separate classifiers: (1) the map classifier, and (2) a field crew. I consider the classification by the field crew to be "error-free" because it is treated as the "true" classification by the user of the map. An accuracy assessment compares the agreement of these two classifiers for the entire thematic map. First, I will discuss the "true" error-matrix, in which each and every pixel in the map is classified by classifiers (1) and (2). Then, I will discuss an estimate of the true error-matrix based on a sample of pixels.

The True Error-Matrix

The accuracy assessment in this example uses a 12 by 12 contingency table, called an "error-matrix" in remote sensing terminology (Congalton, 1991). By convention, each column of the error-matrix represents one of the 12 categories on the map, and each row of the error-matrix represents one of the 12 categories determined by the field crew. Each of the 144 cells of the error-matrix represents the proportion of all pixels in the map that are cross-classified into the corresponding column (i.e., map categories) and row (i.e., true categories). The diagonal of the error-matrix represents the proportion of all pixels for which there is agreement between the map and error-free classifiers, i.e., correct classifications. Scalar statistics, such as kappa and the total proportion of correctly classified pixels, concisely describe the degree of overall agreement between the two classifiers as represented by the 144 cells in this error-matrix.

The total proportion of pixels classified as a certain category in the map equals the sum of the 12 cells within the corresponding column of this error-matrix. The column margin of the error-matrix represents the total proportion of pixels in each category within the thematic map, which corresponds to area statistics that would be produced from this map by a geographic information system. The row margin represents the total proportion of pixels in each category as determined by the error-free classifier (i.e., classifications made by field crew). The difference between the proportions in the row and column margins is called misclassification bias (Czaplewski, 1992a). Thematic maps often overestimate the areal extent of rare categories.

The Estimated Error-Matrix

Although it is practical for the map classifier to assign each and every pixel in the entire map into one of the 12 thematic categories, it is generally impractical for a field crew to classify every pixel. Therefore, the accuracy assessment requires a sample of pixels that are cross-classified by both the map classifier and the error-free classifier. A proper probability sample is sufficient to estimate the error-matrix and make inferences about the true error-matrix. As the sample size increases, the estimated error-matrix will more closely match the true error-matrix.

Simple random sampling of individual pixels is the simplest statistical design for estimation of the error-matrix, and most commercial statistical software can produce valid estimates with this design. However, simple random sampling is usually among the most expensive approaches. More complex sampling designs can reduce costs.

Travel costs for the field crew can be reduced by selecting clusters of pixels that are near each other. Assuming spatial autocorrelation of classification errors, pixels within the same cluster can no longer be considered independent. Most commercial software systems are not designed to correctly estimate an error-matrix with cluster plots.

The cost of reference data can be further reduced by photo-interpretation of reference sites, and classifying a small subset of those sites by field crews. However, photo-interpretation is subject to classification errors, which confounds the assessments of map accuracy (Congalton and Green, 1993). For many years, forest inventories have used double-sampling methods to improve efficiencies of field surveys with photo-interpretation. However, these univariate methods are insufficient to estimate many of the accuracy assessment statistics that are derived from transformations of cells in a multivariate error-matrix.

The following section derives a multivariate estimator for the error-matrix that can accommodate double sampling with cluster plots. Poststratification treats all those pixels classified as a single mapped category as a stratum. The independence among strata improves efficiency. If statistical estimators designed for simple random sampling are applied to reference data from this complex sampling design, then the assessment will be biased and can easily lead to false conclusions (Stehman and Czaplewski, 1998). The objective is to present rigorous derivation of statistical methods for implementation by statisticians who serve users of spatial data. Also, we have implemented these same methods in user-friendly software that we call ACAS (ACcuracy ASsessment software).

METHODS

This section will describe the estimation problem using formal statistical terminology and notation. Next, subpopulations and parameter matrices are defined. Then, I present multivariate sample-survey estimators that use homogeneous sample units, such as independent pixels. These estimators are generalized to deal with clusters of sample units. Next, I introduce the multivariate composite estimator, which I use to combine two independent vector estimates: (1) the estimate from the Phase-1 sample, which includes the imperfect (photo-interpreted) reference classifications, but does not include the error-free classifications (field crew), and (2) the estimate from Phase-2, which includes both field classifications and photo-interpreted classifications. The combination of these two independent estimates produces an estimated error-matrix that uses field classifications as the definition of truth, while using photo-interpreted data to improve efficiency. Then, I present methods to estimate the margin of this error-matrix, which provides unbiased and consistent estimates of areal extent for each thematic category. Finally, I present multivariate methods that transform this error-matrix into accuracy assessment statistics, such as kappa statistics and total proportion of correct classifications. I also derive variance estimators for these statistics.

Description of the Statistical Problem

Consider a thematic map that is comprised of many map units (e.g., pixels or polygons), each of which is imperfectly classified into one of k mutually exclusive categories of land cover by an inexpensive, but fallible, classifier (e.g., remotely sensed data). An "error-free" reference classifier determines the true category for a probability sample of map units or control points (Arbia, 1993). This sample serves as the basis of statistical inference to assess the entire map.

A $k \times k$ contingency table is the basis for most accuracy assessments. The rows represent "error-free" reference classifications, and the columns will represent the imperfect classifications on the map. The ijth element of the contingency table is the estimated joint probability that any map unit is labeled as category j on the map and is truly category i.

An assessment of classification accuracy utilizes various scalar statistics computed from this contingency table, as reviewed by Bishop et al. (1975), Fleiss (1981), and Congalton (1991). The total probability of correct classification is the sum of the diagonal elements of the contingency table. Conditional probabilities of correct classification (e.g., accuracy given that the mapped or error-free classification is a certain category) are the joint probabilities on the diagonal divided by their corresponding row or column marginal probability (Agresti, 1990, p. 9; Green et al., 1993). Weighted, unweighted, and conditional kappa statistics are additional assessment statistics that help (Czaplewski, 1994). One margin of this contingency table provides unbiased estimates of areal extent (Van Deusen, 1994), expressed as proportions of the population. The other margin corresponds to the census, or complete enumeration, of map units, all of which are categorized with the map classifier.

Classifications in the field are typically considered "error-free" reference data. However, field observations are expensive. Photo-interpretation is a less expensive source of reference data, but photo-interpretation errors confound the assessment (Congalton and

Green, 1993). The combination of photo-interpretations with field classifications can improve precision of the assessment statistics and areal estimates using multivariate two-phase sampling.

A thematic map represents a census of all map units, each of which is classified with the map classifier. The $k \times 1$ column margin of the $k \times k$ contingency table can be fixed as a vector of known constants from the census. This reduces the uncertainty in estimates of the k^2 elements within the contingency table and the k elements of the row margin, which correspond to the estimated true proportion of each category.

Czaplewski (1992b) suggested the multivariate composite estimator to produce the desired $k \times k$ contingency table. This uses methods from stochastic processes (Maybeck, 1979, p. 26). The composite estimator combines vector sample estimates from both the Phase-1 photointerpreted sample and the Phase-2 field samples, and fixes one margin of the contingency table through the census of map units and their classifications on the map. Williamson and Haber (1994) review many closely related problems with cross-classified data, but none of the multiple-sample approaches consider margins fixed through a census. The effect of such complex designs can be substantial on tests of hypothesis and estimated confidence intervals (Holt et al., 1980; Rao and Thomas, 1989).

An imperfect but inexpensive map classifier (e.g., digital classification of remotely sensed imagery) assigns a map unit (e.g., a 0.5-ha plot or a 0.1-ha map pixel) to one and only one of k mutually exclusive categories, where k typically ranges between 2 and 30 categories. The population consists of all N map units on the thematic map, where N is known exactly. The true category for any map unit could be determined with an expensive, error-free reference classifier (e.g., field data) that uses the same k categories as the map classifier. An assessment of the accuracy of the map classifier requires the $k \times k$ contingency table $\mathbf{Z}$. The ijth element of $\mathbf{Z}$ is the joint probability that any map unit in the population is classified as category j on the map and is truly category i.

An imperfect classifier (e.g., manual interpretation of aerial photographs) is also available. This imperfect reference classifier will not always agree with the error-free reference classifier. This imperfect reference classifier uses k_y different categories, which can differ from the k categories used by the map and error-free reference classifiers.

The inexpensive map classifier is applied to all N map units in the population or thematic map (e.g., N equals 10^6 to 10^8 units). The expensive, but imperfect, reference classifier can be applied to a small sample of map units (e.g., n_y equals 10^3 to 10^4 units). The more expensive, error-free reference classifier is applied to an even smaller sample of map units (e.g., n_x equals 10^2 to 10^3 units).

The reference data consist of two independent probability samples of map units within the same subpopulation (defined in the next section). The first sample uses map units that are categorized with both the imperfect reference classifier and the map classifier, which is analogous to a first-phase sample. The second sample, which uses the same type of map units, is categorized with all three classifiers: the map, the imperfect reference, and the error-free reference classifiers. This is analogous to a second-phase sample. Each primary sample unit can be an individual map unit, or a cluster of map units. Assume there is negligible locational error (registration error) in location of a map unit in the field (Arbia, 1993, p. 343). The multivariate composite estimator, which is presented in a following section, combines these two samples into a single, more efficient multivariate estimate of the contingency table.

Definitions: Subpopulations and Parameter Matrices

Since the map classification is known for each map unit, the entire population of N map units can be segregated into k subpopulations. The map classification, denoted as category m, is the same for all map units in subpopulation $M = m$. The number of map units in each subpopulation ($N_{M=m}$) is known exactly. Later, independent estimates of k conditional classification probabilities will be made separately for each of the k subpopulations, then transformed and merged into the required $k \times k$ contingency table.

Let $z_{M=m}$ be a $k \times 1$ parameter vector for the subpopulation of map units that are assigned to category m by the map classifier. The ith element of $z_{M=m}$ equals the proportion of map units in this subpopulation that the error-free reference classifier would assign to category I. The sum of all elements in $z_{M=m}$ is exactly 1, and all elements occur in the interval between 0 and 1. The mth column of the $k \times k$ contingency table Z equals $z_{M=m}$ times the prevalence of the mth subpopulation in the total population:

$$Z = \left[z_{M=1}\left(\frac{N_{M=1}}{N}\right) \cdots z_{M=m}\left(\frac{N_{M=m}}{N}\right) \cdots z_{M=k}\left(\frac{N_{M=k}}{N}\right) \right] \tag{1}$$

Next, let $y_{M=m}$ be a $k_y \times 1$ parameter vector for the subpopulation that the map classifier assigned to category m. The ith element of $y_{M=m}$ equals the proportion of all $N_{M=m}$ map units in subpopulation $M = m$ that are assigned to category i, $i \in \{1,...,k_y\}$, by the imperfect reference classifier. The sum of all elements of $y_{M=m}$ equals exactly 1. Furthermore, let the $k_y \times 1$ measurement vector y_p represent the classification outcome for an individual map unit, denoted by subscript p; if the imperfect reference classifier assigns the pth map unit to category i, then the ith element of y_p equals 1 and all other elements equal 0.

Finally, let $X_{M=m}$ be a $k \times k_y$ parameter matrix for subpopulation $M = m$. The ijth element of $X_{M=m}$ is the proportion of map units in this subpopulation that the error-free reference classifier would assign to category i and the imperfect reference classifier would assign to category j. Using the notation of Molina (1989) and Christensen (1991, p. 3), define the Vec as the vectorization operator that stacks columns of a matrix. Let $x_{M=m} = Vec(X_{M=m})$, where $x_{M=m}$ is the equivalent $(kk_y) \times 1$ vector that contains the same parameters as $X_{M=m}$. Rearrangement of the $k \times k_y$ matrix ($X_{M=m}$) into a $(kk_y) \times 1$ stacked vector ($x_{M=m}$) facilitates computation of the $(kk_y) \times (kk_y)$ covariance matrix $\hat{V}x_{M=m}$ for the estimate of $x_{M=m}$. In addition, let the $(kk_y) \times 1$ measurement vector x_p represent the joint classification of a single map unit p. If the error-free reference classifier assigns the pth map unit to category i and the imperfect reference classifier assigns it to category j, then the $[(i-1)k+j]th$ element of x_p equals 1, and all other elements equal 0.

Subpopulation vectors $z_{M=m}$ and $y_{M=m}$ are linear transformations of $x_{M=m}$:

$$y_{M=m} = H_y \, x_{M=m} \tag{2}$$

$$z_{M=m} = H_z \, x_{M=m} \tag{3}$$

H_y and H_z are appropriately structured matrices of zeros and ones, examples of which follow. H_y has dimensions $k_y \times (kk_y)$, and H_z has dimensions $k \times (kk_y)$. These transformations will be exploited by the multivariate composite estimator in a following section.

The following is a simple example. The map and error-free reference classifiers assign map units into $k = 3$ categories: forest (F), nonforest (N), and water (W). The imperfect reference classifier uses $k_y =$ two categories: vegetated (V) and barren (B). Let $(X_{M=m})_{ij}$ represent the ijth element of matrix $\mathbf{X}_{M=m}$; for example, $(X_{M=N})_{FV}$ is the proportion of map units in the subpopulation assigned to the nonforest (N) category by the map classifier that would be classified as forest (F) by the error-free reference classifier and vegetated (V) by the imperfect reference classifier. In this example,

$$\mathbf{X}_{M=N} = \begin{bmatrix} (X_{M=m})_{FV} & (X_{M=m})_{FB} \\ (X_{M=m})_{NV} & (X_{M=m})_{NB} \\ (X_{M=m})_{WV} & (X_{M=m})_{WB} \end{bmatrix} \qquad \mathbf{x}_{M=N} = Vec(\mathbf{X}_{M=N}) = \begin{bmatrix} (X_{M=m})_{FV} \\ (X_{M=m})_{NV} \\ (X_{M=m})_{WV} \\ (X_{M=m})_{FB} \\ (X_{M=m})_{NB} \\ (X_{M=m})_{WB} \end{bmatrix} \tag{4}$$

$$\mathbf{y}_{M=N} = \begin{bmatrix} (X_{M=m})_{FV} & + (X_{M=m})_{NV} & + (X_{M=m})_{WV} \\ (X_{M=m})_{FB} & + (X_{M=m})_{NB} & + (X_{M=m})_{WB} \end{bmatrix}$$

$$= \mathbf{H}_y\, \mathbf{x}_{M=N} = \begin{bmatrix} 1 & 1 & 1 & 0 & 0 & 0 \\ 0 & 0 & 0 & 1 & 1 & 1 \end{bmatrix} \begin{bmatrix} (X_{M=m})_{FV} \\ (X_{M=m})_{NV} \\ (X_{M=m})_{WV} \\ (X_{M=m})_{FB} \\ (X_{M=m})_{NB} \\ (X_{M=m})_{WB} \end{bmatrix} \tag{5}$$

$$\mathbf{Z}_{M=N} = \begin{bmatrix} (X_{M=m})_{FV} & + (X_{M=m})_{FB} \\ (X_{M=m})_{NV} & + (X_{M=m})_{NB} \\ (X_{M=m})_{WV} & + (X_{M=m})_{WB} \end{bmatrix}$$

$$= \mathbf{H}_z\, \mathbf{x}_{M=N} = \begin{bmatrix} 1 & 0 & 0 & 1 & 0 & 0 \\ 0 & 1 & 0 & 0 & 1 & 0 \\ 0 & 0 & 1 & 0 & 0 & 1 \end{bmatrix} \begin{bmatrix} (X_{M=m})_{FV} \\ (X_{M=m})_{NV} \\ (X_{M=m})_{WV} \\ (X_{M=m})_{FB} \\ (X_{M=m})_{NB} \\ (X_{M=m})_{WB} \end{bmatrix} \tag{6}$$

Assume that the pth map unit is classified as nonforest (N) by the map classifier, barren by the imperfect reference classifier (B), and water by the error-free classifier (W). Then this map unit is a member of subpopulation $M = $ N, and its measurement vectors are:

$$\mathbf{y}_p = \begin{bmatrix} 0 \\ 1 \end{bmatrix} \quad \mathbf{x}_p = \begin{bmatrix} 0 \\ 0 \\ 0 \\ 0 \\ 0 \\ 1 \end{bmatrix} \tag{7}$$

Estimates from Homogeneous Sample Units

First, consider a simple random sample that has a fixed sample size of $n_{y|M=m}$ map units from subpopulation $M = m$. Let $S_{y|M=m}$ represent the set of $n_{y|M=m}$ subscripts (p) for the map units in this sample. Each map unit in this sample was classified with the map classifier to determine its subpopulation, and with the imperfect reference classifier (e.g., photo-interpretation). From the previous section, vector $\mathbf{y}_{M=m}$ contains the proportions of map units in this subpopulation that would be classified into each of k_y categories by the imperfect reference classifier, and measurement vector $\mathbf{y}_p$ contains the results of the imperfect reference classifier for the pth map unit (e.g., Equation 7). The vector sample mean provides an efficient and asymptotically unbiased estimate of $\mathbf{y}_{M=m}$ in Equation 2:

$$\hat{\mathbf{y}}_{M=m} = \frac{1}{n_{y|M=m}} \left(\sum_{p \in S_{y|M=m}} \mathbf{y}_p \right) = \mathbf{y}_{M=m} + \mathbf{e}_{y|M=m} \tag{8}$$

where $\mathbf{e}_{y|M=m}$ is the $k_y \times 1$ vector of random sampling errors, and $E[\mathbf{e}_{y|M=m}] = \mathbf{0}$ for a large $n_{y|M=m}$. Equation 8 can be biased for small $n_{y|M=m}$ because zero elements are treated as structural zeros even though they might be sampling zeros (Bishop et al., 1975, Chapters 5 and 12); a structural zero has a probability exactly equal to zero, while a sampling zero element represents a rare event that was not observed in the sample (Bishop et al., 1975, p. 177). Assuming sampling with replacement, the multinomial distribution provides an estimated covariance matrix for these sampling errors ($E[\mathbf{e}_{y|M=m}\mathbf{e}'_{y|M=m}] = \mathbf{V}_{y|M=m}$):

$$\hat{\mathbf{V}}_{y|M=m} = \frac{Diag(\hat{\mathbf{y}}_{M=m}) - \hat{\mathbf{y}}_{M=m}\hat{\mathbf{y}}'_{M=m}}{n_{y|M=m}} \tag{9}$$

where $Diag(\hat{\mathbf{y}}_{M=m})$ is the $k_y \times k_y$ matrix with vector $\hat{\mathbf{y}}_{M=m}$ on the diagonal and all other elements equal zero (Agresti 1990, p. 423). Since the sampling fraction is usually small for thematic maps ($n_{y|M=m}/N_{M=m} < 0.01$), the multinomial distribution will often be a reasonable approximation for sampling without replacement.

A second, independent sample of $n_{x|M=m}$ map units is taken from the same subpopulation ($M = m$). $S_{x|M=m}$ is the set of subscripts (p) for the map units in this second sample. In addition to being classified by the map classifier and the imperfect reference classifier, each map unit in this sample is also classified with the error-free classifier, and the results are represented by measurement vector $\mathbf{x}_p$ (e.g., Equation 7). This provides a sample estimate of $\mathbf{x}_{M=m}$ and its multinomial covariance matrix for sampling errors:

$$\hat{\mathbf{x}}_{M=m} = \sum_{p \in S_{x|M=m}} \frac{\mathbf{x}_p}{n_{x|M=m}} = \mathbf{x}_{M=m} + \mathbf{e}_{x|M=m} \tag{10}$$

$$\hat{\mathbf{V}}_{x|M=m} = \frac{Diag(\hat{\mathbf{x}}_{M=m}) - \hat{\mathbf{x}}_{M=m}\hat{\mathbf{x}}'_{M=m}}{n_{y|M=m}} \tag{11}$$

It is assumed with the multivariate composite estimator that the sampling errors $\mathbf{e}_{y|M=m}$ and $\mathbf{e}_{x|M=m}$ are independent.

Estimates from Clusters of Sample Units

Now consider a cluster of map units as the primary sample unit. Let subscript c denote the cth cluster, and $S_{c|M=m}$ represent the set of $N_{c|M=m}$ subscripts for those map units (p) that make up cluster plot c and are members of subpopulation $M = m$. A map unit can be a member of only one cluster, but more than one subpopulation may occur in a single cluster. The measurement vectors for the cth cluster are:

$$\mathbf{y}_c = \sum_{p \in S_{c|M=m}} \frac{\mathbf{y}_p}{N_{c|M=m}} \tag{12}$$

$$\mathbf{x}_c = \sum_{p \in S_{c|M=m}} \frac{\mathbf{x}_p}{N_{c|M=m}} \tag{13}$$

The ith element of $\mathbf{y}_c$ equals the proportion of category i in the cth cluster, and the sum of all elements in $\mathbf{y}_c$ equals 1. The $[(i-1)k+j]th$ element of $\mathbf{x}_c$ equals the proportion of the cth cluster that is assigned to category i by the error-free reference classifier and category j by the imperfect reference classifier.

The sample mean vectors remain efficient and asymptotically unbiased estimates of $\mathbf{y}$ and $\mathbf{x}$:

$$\hat{\mathbf{y}}_{M=m} = \sum_{c \in S_{y|M=m}} \frac{\mathbf{y}_c}{n_{y|M=m}} = \mathbf{y}_{M=m} + \mathbf{e}_{y|M=m} \tag{14}$$

$$\hat{\mathbf{x}}_{M=m} = \sum_{c \in S_{x|M=m}} \frac{\mathbf{x}_c}{n_{x|M=m}} = \mathbf{x}_{M=m} + \mathbf{e}_{x|M=m} \tag{15}$$

where $S_{y|M=m}$ represents the set of subscripts for the sample of $n_{y|M=m}$ clusters in subpopulation $M = m$ that are measured with the imperfect reference classifier alone, and $S_{x|M=m}$ represents the set of subscripts for sample of the $n_{x|M=m}$ clusters in the same subpopulation that are measured with both the imperfect and error-free reference classifiers. However, the multinomial distribution (Equations 9 and 11) should not be used for the sampling error covariance matrices $\mathbf{V}_y$ and $\mathbf{V}_x$ because sampling errors for map units within the same cluster are not likely to be independent. The sample covariance matrix, which is an asymptotically unbiased moment estimator, is an alternative:

$$\hat{\mathbf{V}}_{y|M=m} = \sum_{c \in S_{y|M=m}} \left[\frac{(\mathbf{y}_c - \hat{\mathbf{y}}_{M=m})(\mathbf{y}_c - \hat{\mathbf{y}}_{M=m})'}{n_{y|M=m} - 1} \right] \tag{16}$$

$$\hat{\mathbf{V}}_{x|M=m} = \sum_{c \in S_{x|M=m}} \left[\frac{(\mathbf{x}_c - \hat{\mathbf{x}})(\mathbf{x}_c - \hat{\mathbf{x}})'}{n_{x|M=m} - 1} \right] \tag{17}$$

Multivariate Composite Estimator

The multivariate composite estimator (Maybeck, 1979, p. 217) combines independent vector estimates $\hat{\mathbf{y}}_{M=m}$ (Equations 8 or 14) and $\hat{\mathbf{x}}_{M=m}$ (Equations 10 or 15) into a more efficient $(kk_y) \times 1$ vector estimate $\hat{\mathbf{x}}_{M=m|y,x}$ for subpopulation $M = m$:

$$\hat{\mathbf{x}}_{M=m|y,x} = (\mathbf{K}_{M=m})\hat{\mathbf{y}}_{M=m} + (\mathbf{I} - \mathbf{K}_{M=m}\mathbf{H}_y)\hat{\mathbf{x}}_{M=m} \tag{18}$$

where $\mathbf{I}$ is the $(kk_y) \times (kk_y)$ identity matrix, $\mathbf{H}_y$ is given in Equations 2 and 5, and $\mathbf{K}_{M=m}$ is defined below. The covariance matrix for this composite estimate ($\hat{\mathbf{V}}_{x|M=m,y,x}$) is:

$$\hat{\mathbf{V}}_{x|M=m,y,x} = \mathbf{K}_{M=m}\hat{\mathbf{V}}_{y|M=m}\mathbf{K}'_{M=m} + (\mathbf{I} - \mathbf{K}_{M=m}\mathbf{H}_y)\hat{\mathbf{V}}_{x|M=m}(\mathbf{I} - \mathbf{K}_{M=m}\mathbf{H}_y)' \tag{19}$$

$$\hat{\mathbf{V}}_{x|M=m,y,x} = (\mathbf{I} - \mathbf{K}_{M=m}\mathbf{H}_y)\hat{\mathbf{V}}_{x|M=m} \tag{20}$$

Covariance matrices $\hat{\mathbf{V}}_{x|M=m}$ and $\hat{\mathbf{V}}_{y|M=m}$ are given in Equations 11 and 17, and 9 and 16, respectively. Equation 19, which is called the "Joseph form" in the Kalman filter, is more numerically reliable than the equivalent expression in Equation 20 (Maybeck, 1979, p. 237).

$\mathbf{K}_{M=m}$ in Equations 18, 19, and 20 is a $(kk_y) \times k_y$ matrix that places optimal weight on each element of the two vector estimates in Equation 18 using the minimum variance criterion. $\mathbf{K}_{M=m}$ is termed the gain matrix in the Kalman filter. It is analogous to the weight in the univariate composite estimator (e.g., see Green and Strawderman, 1986; Gregoire and Walters, 1988), which is inversely proportional to the variances of two a priori scalar estimates. The gain matrix $\mathbf{K}_{M=m}$ for this subpopulation is:

$$\mathbf{K}_{M=m} = \hat{\mathbf{V}}_{x|M=m}\mathbf{H}'_y \left[\mathbf{H}_y\hat{\mathbf{V}}_{x|M=m}\mathbf{H}'_y + \hat{\mathbf{V}}_{y|M=m} \right]^- \tag{21}$$

The bracketed term in Equation 21 is a singular covariance matrix. Therefore, Equation 21 uses the generalized inverse for a symmetric matrix (Graybill, 1969, pp. 113–115).

The $(kk_y) \times 1$ composite estimate $\hat{\mathbf{x}}_{M=m|y,x}$ for subpopulation $M = m$ (Equation 18) is the basis for the mth column in the estimated $k \times k$ contingency table $\hat{\mathbf{Z}}$, and is related to the $k \times 1$ vector $\mathbf{z}_{M=m}(N_{M=m}/N)$ in Equation 1. Estimate $\hat{\mathbf{z}}_{M=m}$ is a vector of proportions that sums to exactly 1, and it represents the estimated conditional probabilities of a map unit being assigned to any one of the k categories by the error-free reference classifier, given that the map unit is assigned to category m by the map classifier. $\hat{\mathbf{z}}_{M=m}$ is a linear transformation of $\hat{\mathbf{x}}_{M=m|y,x}$ (Equations 3 and 6), with its corresponding $k \times k$ covariance matrix $\hat{\mathbf{V}}_{z|M=m}$:

$$\hat{\mathbf{z}}_{M=m} = \mathbf{H}_z \, \hat{\mathbf{x}}_{M=m|y,x} \tag{22}$$

$$\hat{\mathbf{V}}_{z|M=m} = \mathbf{H}_z \, \hat{\mathbf{V}}_{x|M=m,y,x} \, \mathbf{H}_z' \tag{23}$$

The multivariate composite estimator (Equations 18 to 23) is applied k different times, once for each map category m; then, vector estimates $\hat{\mathbf{z}}_{M=m}$ are concatenated to form the estimated $k \times k$ contingency table $\hat{\mathbf{Z}}$ (Equation 1).

Let $Vec(\hat{\mathbf{Z}})$ be the $k^2 \times 1$ vectorized version of $\hat{\mathbf{Z}}$:

$$Vec(\hat{\mathbf{Z}}) = \begin{bmatrix} \hat{\mathbf{z}}_{M=1}(N_{M=1} / N) \\ \hat{\mathbf{z}}_{M=m}(N_{M=m} / N) \\ \hat{\mathbf{z}}_{M=k}(N_{M=k} / N) \end{bmatrix} \tag{24}$$

Assume that sampling errors for each subpopulation $M = m$ are independent of the sampling errors for all other subpopulations by design. In this case, the estimation error covariance matrix for $Vec(\hat{\mathbf{Z}})$ is the $k^2 \times k^2$ matrix with submatrices $\hat{\mathbf{V}}_{z|M=m}$ (Equation 23) on the diagonal, and all other elements equal 0:

$$\hat{\mathbf{V}}_{Vec(\hat{\mathbf{Z}})} = \begin{bmatrix} \hat{\mathbf{V}}_{z|M=1} & 0 & 0 \\ 0 & \hat{\mathbf{V}}_{z|M=m}\left(\dfrac{N_{M=m}}{N}\right)^2 & 0 \\ 0 & 0 & \hat{\mathbf{V}}_{z|M=k}\left(\dfrac{N_{M=k}}{N}\right)^2 \end{bmatrix} \tag{25}$$

This covariance matrix will be used to estimate standard errors of statistics that are functions of the estimated contingency table $\hat{\mathbf{Z}}$.

The multivariate composite estimator cannot always be used to estimate $\hat{\mathbf{z}}_{M=m}$ for subpopulation $M = m$ (Equations 18, 22, and 24). If the error-free classifier assigns all sample units in this subpopulation to only one category, say i, then the ith element of $\hat{\mathbf{z}}_{M=m}$ equals exactly 1, and $\hat{\mathbf{V}}_{x|M=m} = \mathbf{0}$. Alternatively, the sample estimate $\hat{\mathbf{y}}_{M=m}$ might not exist for this subpopulation; the association between the imperfect and error-free classifications might be very poor; or the imperfect classifier might assign all sample units in this subpopulation to only one category. If these situations occur, and the sample estimate with the error-free reference classifications (Equations 10 or 15) exists for this subpopulation, then Equations 22 and 23 for the composite estimator can be replaced by $\hat{\mathbf{z}}_{M=m} = \mathbf{H}_z \, \hat{\mathbf{x}}_{M=m}$ and $\hat{\mathbf{V}}_{z|M=m} = \mathbf{H}_z \, \hat{\mathbf{V}}_{x|M=m} \, \mathbf{H}'_z$ (see Equations 10, 11, 15, and 17), and used in Equations 24 and 25 to estimate the mth column of the contingency table. (This approach also satisfies situations in which $\hat{\mathbf{y}}_{M=m}$ does not exist by design because the imperfect reference classifier is omitted. This permits estimation of the contingency table with heterogeneous cluster plots rather than the typical homogeneous plots, which are classified into one and only one category by each classifier.) If the sample estimate $\hat{\mathbf{x}}_{M=m}$ does not exist, then no estimate for subpopulation $M = m$ is possible, and the mth column of the contingency table will be missing.

The multivariate composite estimator is vulnerable to numerical errors. This problem is common whenever random errors for one vector estimate are much greater than random errors for another, e.g., $det(\mathbf{H}_y \, \hat{\mathbf{V}}_{x|M=m} \, \mathbf{H}'_y) \gg det(\hat{\mathbf{V}}_{y|M=m})$, which can occur when sample sizes $n_{x|M=m}$ and $n_{x|M=m}$ are very different in Equations 8 to 17. Results should always be scrutinized for symptoms of numerical problems, such as: vectors of estimated proportions that do not sum to 1; corresponding covariance matrices that do not sum to 0; negative elements on the diagonal of any covariance matrix; or asymmetric covariance matrices. Solutions include the following:

- All computational routines (e.g., generalized inverse) should use maximal numerical precision and be robust to numerical problems.
- The "Joseph form" for $\hat{\mathbf{V}}_{x|M=m,y,x}$ in Equation 19 should be used rather than Equation 20 because the former is better conditioned and more numerically reliable (Maybeck, 1979, p. 237).
- Matrix dimensions can be reduced by eliminating rows of zeros and their corresponding columns in $(\mathbf{H}_y \, \hat{\mathbf{V}}_{x|M=m} \mathbf{H}'_y + \hat{\mathbf{V}}_{y|M=m})$ in Equation 21, although this requires careful bookkeeping in order to expand the resulting composite estimate ($\hat{\mathbf{x}}_{M=m|y,x}$ in Equation 18) back to the original structure of $\hat{\mathbf{x}}_{M=m}$.
- Composite estimation algorithms that use square roots of the covariance matrices often solve stubborn numerical problems (Maybeck, 1979, pp. 377–391).

Validation methods are especially important in composite estimation with remotely sensed data, where logistic and technological difficulties often breed subtle procedural aberrations. Maybeck (1979, p. 229) shows that the vector of residual differences ($\mathbf{H}_y \hat{\mathbf{x}}_{M=m} - \hat{\mathbf{y}}_{M=m}$) has an expected covariance matrix ($\mathbf{H}_y \hat{\mathbf{V}}_{x|M=m} \mathbf{H}'_y + \hat{\mathbf{V}}_{y|M=m}$). Each element of this residual vector for each subpopulation m should be tested for bias, assuming the estimation errors are normally distributed with zero mean and variance on the diagonal of the estimated covariance matrix. Several suspiciously large residuals indicate potential procedural errors. However, off-diagonal elements of the covariance matrix are not necessarily zero, and these separate tests are not mutually independent for the same subpopulation. Therefore, multiple residuals for a subpopulation can be standardized so that they are expected to be independent and identically distributed. First, the dimensions of the residual vector and its covariance matrix are reduced to achieve full rank, then standardized residuals ($\mathbf{r}_{M=m}$) are computed with the Cholesky square root of its covariance matrix:

$$\mathbf{r}_{M=m} = \left[\left(\mathbf{H}_y \hat{\mathbf{V}}_{x|M=m} \mathbf{H}_y' + \hat{\mathbf{V}}_{y|M=m} \right)^{1/2} \right]^{-1} \left(\mathbf{H}_y \hat{\mathbf{x}}_{M=m} - \hat{\mathbf{y}}_{M=m} \right) \tag{26}$$

where $\mathrm{E}[\mathbf{r}_{M=m}] = \mathbf{0}$ and $\mathrm{E}[\mathbf{r}_{M=m} \mathbf{r}'_{M=m}] = \mathbf{I}$. If all sampling and estimation assumptions are valid, then the standardized residuals from all subpopulations can be pooled, and their pooled mean and variance have expected values of 0 and 1, respectively. These expectations are validated with a t-test for 0 mean with variance 1, and a χ^2 test for variance equal to 1 (see Hoel, 1984, pp. 140–143, 281–284, 298–300). If the validation tests cast doubt upon these expectations, then possible procedural problems should be investigated.

Estimates of Areal Extent

Environmental evaluations and forecasting models often require statistical tabulation of the area occupied by each cover category. These areal estimates are available through enumeration of map units that are classified into the k categories by the imperfect map classifier. However, misclassification can make this enumeration a biased estimate of the true areal extent, especially for rare cover types (Czaplewski, 1992a). Czaplewski and Catts (1992) and Walsh and Burk (1993) reviewed the literature that considers calibration for misclassification bias. However, they did not discuss unbiased areal estimates that use the row margin of the estimated contingency table $\hat{\mathbf{Z}}$ (Equation 24). The ijth element of $\mathbf{Z}$, denoted by Z_{ij}, is the proportion of map units in the population that are classified as category j by the map classifier and truly are category i; therefore, the proportion of map units that are truly category i (p_i) equals:

$$p_{i\cdot} = \sum_{j=1}^{k} Z_{ij} \tag{27}$$

To estimate the true areal extent of each category in the absence of misclassification, the estimated contingency table ($\hat{\mathbf{Z}}$) is substituted for the unknown true matrix in Equation 27, which provides an asymptotically unbiased estimator (Molina C., 1989, p. 122). Since estimation errors for elements of $\hat{\mathbf{Z}}$ are independent between sub populations, which are defined by the imperfect map classifier and denoted by subscript j in Equation 27, the variance for $p_{i\cdot}$, denoted $\hat{V}_{\hat{p}_{i\cdot}}$, is simply the sum of the variances for each Z_{ij} on the diagonal of covariance matrix $\hat{\mathbf{V}}_{Vec(z)}$ in Equation 25.

Matrix algebra provides a concise formulation for areal estimates and their covariance matrix. Let the $k \times 1$ vector $\hat{\mathbf{p}}_I$ represent the estimated true proportions of each of the k categories, *i.e.,* the row margin of $\hat{\mathbf{Z}}$; $\hat{\mathbf{p}}_I$ is a linear transformation of $\hat{\mathbf{Z}}$:

$$\hat{\mathbf{p}}_{i\cdot} = \mathbf{D}'_{\mathbf{p}_{i\cdot}} Vec(\hat{\mathbf{Z}}) \tag{28}$$

where $\mathbf{Dp}_r$ is the following $k^2 \times k$ matrix of zeros and ones:

$$\mathbf{D}_{\mathbf{p}_{i\cdot}} = \begin{bmatrix} \mathbf{I} \\ \mathbf{I} \\ \vdots \\ \mathbf{I} \end{bmatrix} \tag{29}$$

$\mathbf{I}$ is the $k \times k$ identity matrix. The covariance matrix for random estimation errors in $\hat{\mathbf{p}}_I$ is the corresponding linear transformation of the $k^2 \times k^2$ covariance matrix for $Vec(\hat{\mathbf{Z}})$ in Equation 25:

$$\hat{\mathbf{V}}_{\mathbf{p}_{i\cdot}} = \mathbf{D}'_{\mathbf{p}_{i\cdot}} \, \hat{\mathbf{V}}_{Vec(\hat{Z})} \, \mathbf{D}_{\mathbf{p}_{i\cdot}} \tag{30}$$

Environmental evaluations can require confidence intervals for areal estimates. Confidence intervals often use the normal distribution with an estimated mean and variance. However, the normal distribution is unrealistic for proportions that are near 0 or 1, where the binomial distribution is a more reasonable assumption. The parameters of the binomial distribution are the number of independent trials (n) and the estimated probability of success ($\hat{p}$), with estimated variance $\hat{V}_{\hat{p}} = \hat{p}(1-\hat{p})/n$. The number of trials ($n$), in the context of a parameter of the binomial distribution, does not pertain to composite estimates; however, the composite estimator does provide estimates $\hat{p}$ and $\hat{V}_{\hat{p}}$. Based on the method of moments approach that Brier (1980) used with cluster sampling for the Dirichlet-multinomial distribution, $\hat{p}$ and $\hat{V}_{\hat{p}}$ provide an estimate of parameter n for the binomial distribution:

$$\hat{n} = \frac{\hat{p}(1-\hat{p})}{\hat{V}_{\hat{p}}} \tag{31}$$

The approximate confidence bounds ($\hat{p}_{LO}$ and $\hat{p}_{UP}$) for $\hat{p}$ may be estimated as follows (Rothman, 1986, p. 167). Let a represent an arbitrary confidence level, e.g., $\alpha = 0.95$, and $P(K \geq \hat{p}\hat{n})$ represent the probability that ($\hat{p}\hat{n}$) or more successes are observed out of $\hat{n}$ Bernoulli trials, where ($\hat{p}\hat{n}$) and $\hat{n}$ are rounded to the nearest integer. The lower confidence bound is the probability of success ($\hat{p}_{LO}$) with the binomial distribution for which $P(K \geq \hat{p}\hat{n}) = (1-\alpha)/2$, and the upper bound is the $\hat{p}_{UP}$ for which $P(K \leq \hat{p}\hat{n}) = (1-a)/2$. Solution is by bisection. Confidence intervals so computed are always bounded by 0 and 1. Except when $\hat{p}$ is near 0 or 1, these confidence intervals are approximately equal to intervals that assume a normal distribution for a sufficiently large sample size.

Accuracy Assessment Statistics and Variances

The most common statistic used to assess accuracy is the total proportion of correctly classified map units (p_o), called overall accuracy by Congalton (1991). Once the contingency table is estimated (Equations 22 and 24), the estimated overall accuracy ($\hat{p}_o$) simply equals the sum of the diagonal elements of $\hat{\mathbf{Z}}$. The variance for the overall accuracy statistic $\hat{V}_{\hat{p}_o}$ is the sum of the diagonal elements of $\hat{\mathbf{V}}_{Vec(Z)}$ (Equations 23 and 25) that correspond to diagonal elements of $\hat{\mathbf{Z}}$. This is expressed in matrix algebra as a prelude to formulae for more complex accuracy statistics:

$$\hat{p}_o = \mathbf{d}'_{p_o} Vec(\hat{\mathbf{Z}}) \tag{32}$$

$$\hat{V}_{\hat{p}_o} = \mathbf{d}'_{p_o} \hat{\mathbf{V}}_{Vec(\hat{Z})} \mathbf{d}_{p_o} \tag{33}$$

where $\mathbf{d}p_c$ is the $k^2 \times 1$ vector in which the $[(i-1)k+i]$th elements equal 1 ($i = 1,2,...,k$), and all other elements equal 0. Confidence intervals are approximated with $\hat{p}_o$, $\hat{V}\hat{p}_c$, and the binomial assumption (Equation 31).

The expected probabilities of correct classification for a category through chance agreement equals the product of the row and column margins for that category in the contingency table. This hypothesis is the basis for Cohen's kappa test, which is commonly applied in remote sensing studies (Congalton, 1991). The column margin for $\hat{\mathbf{Z}}$ is known exactly through the census of map units, each of which is categorized by the map classifier. Let $k \times 1$ vector $\mathbf{p}_{\cdot j}$ represent the column margin of $\hat{\mathbf{Z}}$, in which the ith element equals the constant $N_{M=1}/N$ (Equation 1). The estimated expected accuracy under chance agreement $\hat{p}_o$ is the product of the fixed column ($\mathbf{p}_{\cdot j}$) and the estimated row margin ($\hat{\mathbf{p}}_{i\cdot}$ in Equation 28), with covariance matrix $\hat{V}_{\hat{p}_c}$:

$$\hat{p}_c = \mathbf{p}'_{\cdot j} \hat{\mathbf{p}}_{i\cdot} = \mathbf{p}'_{\cdot j} \mathbf{D}'_{p_{i\cdot}} Vec(\hat{\mathbf{Z}}) = \mathbf{d}'_{p_c} Vec(\hat{\mathbf{Z}}) \tag{34}$$

$$\hat{V}_{\hat{p}_c} = \mathbf{d}'_{p_c} \hat{\mathbf{V}}_{Vec(\hat{Z})} \mathbf{d}_{p_c} \tag{35}$$

$$\mathbf{d}'_{p_c} = \mathbf{p}'_{\cdot j} \mathbf{D}'_{p_{i\cdot}} = \left[\mathbf{p}'_{\cdot j} | \mathbf{p}'_{\cdot j} | \quad | \mathbf{p}'_{\cdot j} \right] \tag{36}$$

This suggests a test of hypothesis that the overall classification accuracy is no greater than that expected by chance, i.e., the difference between the observed overall accuracy ($\hat{p}_o$ in Equation 32) and chance accuracy ($\hat{p}_c$ in Equation 34) equals zero:

$$\hat{p}_o - \hat{p}_c = [1,-1] \begin{bmatrix} \hat{p}_o \\ \hat{p}_c \end{bmatrix} = [1,-1] \begin{bmatrix} \mathbf{d}'_{p_o} \\ \mathbf{d}'_{p_c} \end{bmatrix} Vec(\hat{\mathbf{Z}}) \tag{37}$$

$$\hat{V}_{\hat{p}_o - \hat{p}_c} = [1,-1] \begin{bmatrix} \mathbf{d}'_{p_o} \\ \mathbf{d}'_{p_c} \end{bmatrix} \hat{\mathbf{V}}_{Vec(\hat{Z})} \begin{bmatrix} \mathbf{d}'_{p_o} \\ \mathbf{d}'_{p_c} \end{bmatrix} [1,-1] \tag{38}$$

The normal distribution with mean 0 and variance $\hat{V}\hat{p}_o - \hat{p}_c$ provides an approximate probability that the null hypothesis is true, although the limiting distribution of this statistic has not been established. Czaplewski (1994) provides a Taylor series approximation for covariance matrix $\hat{V}_{Vec(Z)}$ in Equation 38 that is more compatible with the null hypothesis in this test.

Some land cover types are very accurately categorized by a map classifier, while others are less successfully classified. The overall accuracy statistic does not isolate differences among individual categories. However, conditional probabilities of correct classification concisely and intuitively describe these differences (e.g., Fleiss, 1981, p. 214). Green et al. (1993) and Stehman (1996) developed variance estimators for these statistics when prestratification is based on the remotely sensed classification, there is simple random sampling within strata, and sample units are homogeneous so that the multinomial distribution applies. The following two paragraphs provide more general results for estimates of conditional probabilities from double sampling.

Consider the conditional probability of correct classification given that the map classifier assigns a map unit to category $M = m$. This is also called "user's accuracy" in the remote sensing and quality control literatures (Congalton, 1991). The vector $\hat{z}_{M=m}$ in Equation 22 is the transformed composite estimate of the conditional probabilities given that the map classifier assigned the map unit to the mth category (i.e., subpopulation $M = m$). Therefore, user's accuracy for category m is simply the mth element of $\hat{z}_{M=m}$, with an estimated variance equal to the mth diagonal element of $\hat{V}_{z|M=m}$ in Equation 23. The confidence interval for user's accuracy for the mth category is approximated with the binomial distribution (Equation 31).

The conditional probability of correct classification, given that the error-free reference classification is category m, is called "producer's accuracy" (Congalton, 1991). The $k \times 1$ vector of producer's accuracies, $\hat{p}_{(i|i\cdot)}$, equals the $k \times 1$ diagonal of the $k \times k$ contingency table, $diag(\hat{Z})$, divided by its row margin (Equation 28). The covariance matrix for producer's accuracies is more difficult to estimate than the covariance matrix for user's accuracies because estimation errors among rows of $\hat{z}_{M=m}$ are not independent (Equation 25). Czaplewski (1994) derives a Taylor series approximation for this situation. First, define the $k \times k$ matrix $H_{p(i\cdot)}$, in which all off-diagonal elements are zero, and all diagonal elements equal the iith element of $\hat{Z}$ divided by the squared inverse of the ith element of $\hat{p}_I$, $i \in \{1,\cdots,k\}$, in Equation 28. Next, define the $k \times k$ matrix G_i, $I \in \{1,\cdots,k\}$, in which all elements are zero except the iith element, which equals the inverse of the ith element of $(\hat{p}_I)$. Finally, define the $k^2 \times k$ matrix $D_{(i|i\cdot)}$ as:

$$D_{(i|i\cdot)} = \begin{bmatrix} G_1 \\ \vdots \\ G_m \\ \vdots \\ G_k \end{bmatrix} - \begin{bmatrix} H_{p(i\cdot)} \\ \vdots \\ H_{p(i\cdot)} \\ \vdots \\ H_{p(i\cdot)} \end{bmatrix} \tag{39}$$

The approximate $k \times k$ covariance matrix for the $k \times 1$ vector of estimated producer's accuracies ($\hat{p}_{(III\cdot)}$) equals:

$$\hat{\mathbf{V}}_{\hat{\mathbf{p}}_{(ili\cdot)}} = \mathbf{D}'_{(ili\cdot)} \hat{\mathbf{V}}_{Vec(\hat{\mathbf{Z}})} \mathbf{D}_{(ili\cdot)} \tag{40}$$

Czaplewski (1994) provides examples of the matrix in Equation 40. The confidence intervals for elements of $\hat{\mathbf{p}}_{(III\cdot)}$ (producer's accuracies) are approximated with their corresponding diagonal elements of covariance matrix $\hat{\mathbf{V}}\hat{\mathbf{p}}_{(III\cdot)}$ and the binomial distribution (Equation 31). In addition, Czaplewski (1994) formulates a test of the hypothesis that individual user's and producer's accuracies do not differ from what is expected by chance.

The variances in Equations 35 and 38, and the covariance matrix for user's accuracies, assume that the column margin $\mathbf{p}_{\cdot j}$ is a vector of known constants. This is true when the map classifier is applied to all members of the population, i.e., the column margin of $\hat{\mathbf{Z}}$ is fixed through a census of all pixels. However, these equations do not apply to more general situations, which are considered by Czaplewski (1994).

The coefficient of agreement (kappa) also quantifies overall accuracy in a contingency table $\hat{\mathbf{Z}}$ relative to that expected by chance (Cohen, 1960, 1968). This statistic is commonly used in remote sensing (Congalton, 1991). It is sometimes weighted to consider partial agreement. Let $\mathbf{W}$ represent the $k{\times}k$ matrix of constants in which the ijth element (w_{ij}) is the weight or "partial credit" chosen by the user for the agreement when a map unit is classified as category j by the imperfect map classifier and category i by the error-free reference classifier. The unweighted kappa statistic is merely a special case of the weighted kappa, in which $\mathbf{W} = \mathbf{I}$, the identity matrix. Let $\hat{p}_{o|W} = \mathbf{1}'[\mathbf{W}{*}\hat{\mathbf{Z}}]\mathbf{1}$ represent the weighted overall accuracy observed in the sample, and $\hat{p}_{c|W} = \mathbf{1}'[\mathbf{W}{*}(\hat{\mathbf{p}}_I.\mathbf{p}'_{\cdot j})]\mathbf{1}$ (see Equations 28 and 34) represent the corresponding accuracy expected by chance, where $\mathbf{1}$ is the $k{\times}1$ vector of 1's and $*$ represents element-by-element multiplication (i.e., the ijth element of $\mathbf{A}{*}\mathbf{B}$ is $a_{ij}b_{ij}$). The estimated weighted kappa statistic ($\hat{\kappa}_w$) is:

$$\kappa_w = \frac{\hat{p}_{o|W} - \hat{p}_{c|W}}{1 - \hat{p}_{c|W}} \tag{41}$$

Cohen derived a variance approximation for kappa in the special case of simple random sampling, where each unit is classified into one and only one category, and Stehman (1996, 1997) has developed variance approximations for stratified random sampling. However, these do not apply to the multivariate composite estimator. Czaplewski (1994) derives a Taylor-series variance approximation for the weighted kappa statistic (k_w) in the more general case, where the contingency table $\hat{\mathbf{Z}}$ is estimated with any appropriate method that provides covariance matrix $\hat{\mathbf{V}}_{Vec(Z)}$ (e.g., Equation 25). First, define the $k{\times}1$ vectors $\mathbf{w}_{i\cdot} = (\mathbf{W}\,\mathbf{p}_{\cdot j})$ from Equation 34, and $\hat{\mathbf{w}}_{\cdot j} = (\mathbf{W}'\hat{\mathbf{p}}_I.)$ from Equation 28, and $k^2{\times}1$ vector $\mathbf{d}_k$:

$$\mathbf{d}_\kappa = \left[\frac{Vec(\mathbf{W})}{(1-\hat{p}_{c|W})} - \frac{(1-\hat{p}_{o|W})}{(1-\hat{p}_{c|W})^2}\right]\left\{\begin{bmatrix}\mathbf{w}_{1\cdot}\\[4pt]\mathbf{w}_{m\cdot}\\[4pt]\mathbf{w}_{k\cdot}\end{bmatrix} + Vec\begin{bmatrix}\hat{\mathbf{w}}'_{\cdot 1}\\[4pt]\hat{\mathbf{w}}'_{\cdot m}\\[4pt]\hat{\mathbf{w}}'_{\cdot k}\end{bmatrix}\right\} \tag{42}$$

where $Vec(\mathbf{W})$ is the $k^2 \times 1$ vector version of the $k \times k$ weighting matrix $\mathbf{W}$, and the Vec operator is defined for $\mathbf{x}_{M=m}$ in Equation 2; Czaplewski (1994) provides examples. Next, the approximate variance of $\hat{\kappa}_w$ is estimated as:

$$\hat{V}_{\hat{\kappa}_w} = \mathbf{d}'_\kappa \, \hat{V}_{Vec(\hat{\mathbf{Z}})} \, \mathbf{d}_\kappa \tag{43}$$

The estimated kappa and its variance (Equations 41 and 43) are used with the normal distribution to estimate confidence intervals and test hypotheses. Czaplewski (1994) derives other variance approximations for the weighted and unweighted kappa statistics under the null hypothesis of chance agreement, and the conditional kappa statistics (Light, 1971) for the rows and columns of the contingency table.

Equations 30, 33, 35, 40, and 43 represent variance estimators for areal extent and statistics that assess accuracy of classifications on the map. They are formulated as vector operators on the covariance matrix $\hat{V}_{Vec(z)}$, which is highly partitioned (Equation 25) because of the assumed independence of sampling errors among subpopulations. This partitioned structure can reduce numerical errors and matrix dimensions. Let the following represent this general structure:

$$\begin{bmatrix} \mathbf{D}'_1 \cdots \mathbf{D}'_m \cdots \mathbf{D}'_k \end{bmatrix} \begin{bmatrix} \mathbf{V}_1 & \cdots & \mathbf{0} & \cdots & \mathbf{0} \\ \mathbf{0} & \cdots & \mathbf{V}_m & \cdots & \mathbf{0} \\ \mathbf{0} & \cdots & \mathbf{0} & \cdots & \mathbf{V}_k \end{bmatrix} \begin{bmatrix} \mathbf{D}_1 \\ \mathbf{D}_m \\ \mathbf{D}_k \end{bmatrix} = \sum_{m=1}^{k} \mathbf{D}'_m \mathbf{V}_m \mathbf{D}_m \tag{44}$$

The left-hand side of Equation 44 contains a $k^2 \times k^2$ covariance matrix; however, the right-hand side contains $k \times k$ covariance matrices. This is very important for detailed classification systems. For example, if the number of categories is $k = 20$, then the covariance matrix on the left-hand side has dimensions 8000×8000, while those on the right-hand side have dimensions 400×400.

DISCUSSION

Classification systems frequently have detailed categories to meet the needs of particular analyses, but accuracy of map classifiers is typically poor for detailed classification systems. The classifier often confuses certain similar categories, and the number of similar categories increases as the classification system becomes more detailed. While increased thematic resolution provides more information about spatial patterns, increased resolution makes modeling of those patterns more difficult (Costanza and Maxwell, 1994). Therefore, the map analyst must often simplify the classification system to attain maps that have reasonable classification accuracy. Statistical assessment of classification accuracy naturally leads to more informed decisions regarding these simplifications. Conditional probabilities of correct classification (user's and producer's accuracies for each category) help strike a compromise between analysts' applications and the classification errors that are endemic to thematic mapping. These compromises are facilitated with user-friendly software, copies of which are available from the author. Other software systems are capable of generating the necessary estimates, especially systems with a matrix language and a library of linear algebra routines.

Occasionally, random errors severely distort estimates of accuracy. Gains in efficiency from the imperfect reference classifier (e.g., photo-interpretation) reduce risk of incorrect evaluations of map reliability. For this reason, reasonable confidence intervals and variance estimates for accuracy statistics are crucial for prudent use of thematic maps. Estimates of classification accuracy improve as registration accuracy increases, as sample sizes for estimates $\hat{y}_{M=m}$ and $\hat{x}_{M=m}$ increase (Equations 8 to 17), and as accuracy of the imperfect reference classifier increases. The latter might improve with reductions in classification detail (k_y in Equation 2). All these components affect the cost and reliability of estimating areal extent and accuracy assessments of thematic maps.

Spatial analyses with geographic information systems often generate areal estimates from a small portion of a thematic map, but these estimates can include substantial misclassification. Multivariate calibration methods can correct for this bias (Tenenbein, 1972; Czaplewski and Catts, 1992; Walsh and Burk, 1993) with a transformation of the contingency table ($\hat{Z}$ from Equations 1 and 24). However, the calibrated areal estimates do not identify the spatial location of classification errors. Also, the calibration model assumes that misclassification probabilities are the same for all portions of the thematic map, which might not be reasonable for small-area estimates. For example, mountain shadows often increase classification errors with multispectral satellite data. The contingency table incorporates this source of error in proportion to the amount of shadow within the entire map. However, a portion of the map can have a very different proportion of shadow, and thus, very different misclassification probabilities. Bauer et al. (1994) propose a solution to this problem, in which a univariate composite method combines predictions from local and global calibration models.

The multivariate composite approach can be expanded to multiway contingency tables. This permits hierarchical loglinear models, and related logit models, and associated methods for systematic testing of hypotheses, similar to analysis of variance for continuous data (Rao and Thomas, 1989; Molina C., 1989). Such methods permit testing hypotheses related to causes of classification errors, and possibly other hypotheses. However, the methods used in this chapter avoid logarithmic transformations for estimation because retransformation bias is very problematic, especially for proportions near zero.

The methods presented here include cluster sampling without the imperfect reference classifier as a special case. This efficient estimator is an alternative to the univariate methods of Stehman (1997). In addition, the difference between the realized overall accuracy and the overall accuracy expected by chance alone (Equation 37) is a new indicator of classification accuracy; Equation 38 provides the variance estimator for this statistic under prestratification. Czaplewski (1994) derives their variance estimators. These statistics provide a more reasonable null hypothesis (Stehman, 1997), and are more easily interpreted than Cohen's kappa and partial kappa statistics.

The methods proposed above also apply to situations where the imperfect reference classifier is not photo-interpretation. For example, the imperfect reference data might represent sample units from a different monitoring program, where inconsistencies in protocol and definitions cause "imperfect" classifications, or the imperfect reference data might come from an old survey, where recent changes in land cover cause previous classifications to be imperfect. Lastly, the reference classifications might be considered imperfect if they come from plots that are not well registered between the satellite images and their field locations. The cost of accurate registration could be restricted to a subsample of "error-free" plots.

The methods in this chapter assume each sample unit has the same configuration for the map classifier, the imperfect reference classifier, and error-free reference classifier, which is similar to sample units in a two-phase sampling design. A cluster of pixels in a 1-ha field plot is an example. However, photo-interpretation of larger sample units often adds little marginal cost (Czaplewski and Catts, 1988; Bauer et al., 1994), as in a two-stage sample design. The multivariate composite estimator can accommodate two-stage designs, but requires a different formulation from that considered here.

The composite estimator assumes independence of sampling errors among subpopulations. This assumption is suspect for poststratification by subpopulation of a simple random sample because sample sizes for each subpopulation are not fixed by design. This assumption is also suspect if a cluster plot contains map units from more than one subpopulation. Practical considerations make simple random sampling and heterogeneous cluster plots important options in assessing accuracy. Although the independence assumption is not strictly required by the composite estimator (Czaplewski, 1992a), the independence assumption does reduce numerical problems (e.g., Equation 44). The bootstrap estimator exploits the numerical advantages of this independence assumption, and accounts for dependent sampling errors through resampling. However, bootstrap variance estimates might require hours of computation time, compared with seconds for composite variance estimates. Therefore, the composite estimator can provide reasonable preliminary estimates, especially when assessments of accuracy are used to help simplify the classification system, while final estimates are made with bootstrap methods.

The composite estimator ($\hat{\mathbf{Z}}$ from Equations 18 to 24) is the unbiased minimum variance estimator if $\hat{\mathbf{y}}_{M=m}$ and $\hat{\mathbf{x}}_{M=m}$ are unbiased and the covariance matrices $\mathbf{V}_x$ and $\mathbf{V}_y$ are known. However, only the estimates $\hat{\mathbf{V}}_x$ and $\hat{\mathbf{V}}_y$ are usually known; the composite estimator remains unbiased, but will be suboptimal. Since $\hat{\mathbf{Z}}$ is asymptotically unbiased, and the statistics used to estimate areal extent and assess accuracy are functions of the k^2 estimates in $\hat{\mathbf{Z}}$, these estimates will also be asymptotically unbiased (Molina C., 1989, p. 122; Särndal et al., 1992, p. 168). However, vector estimates $\hat{\mathbf{y}}_{M=m}$ and $\hat{\mathbf{x}}_{M=m}$ in Equations 8 to 17 can be biased for small sample sizes because each zero element in $\hat{\mathbf{y}}_{M=m}$ and $\hat{\mathbf{x}}_{M=m}$ is treated as though the true probability is exactly zero (Bishop et al., 1975, p. 177). A zero element might represent a rare event that was not observed in the sample (sampling zero), even though the true probability of the event exceeds zero. This can bias the composite estimator. Deficiencies in sampling design and measurement protocols can introduce additional biases. The verification techniques described earlier in the Methods section should always be used to detect subtle inconsistencies in the data collection, processing, and estimation processes.

A reference classifier is rarely free of all classification error. Errors are introduced into reference data by errors in locating sample plots on the ground and on the remotely sensed imagery, changes in plot condition over time, measurement and recording errors, and within-plot sampling errors. However, accuracy assessments treat the highest resolution reference data as though they were perfect. The term "error-free reference classifier" emphasizes this assumption to the user of an accuracy assessment. The user should thoroughly understand how the reference data were gathered, and condition their interpretation of the accuracy assessment based on this understanding.

CONCLUSIONS

The contingency table is sufficient for assessing accuracy and estimating the area of different categories of land cover, while maintaining compatibility between these two objectives. The multivariate composite estimator provides efficient estimates of the contingency table because it combines multiple sources of data with an asymptotically unbiased, minimum variance approach. The multivariate composite estimator is a relatively simple method, which permits use of more complex sampling designs and sample units. With statistical methods presented in this chapter, the complexity of the sampling design and estimation method might no longer be a major problem. Rather, the remaining problems might be more pragmatic, such as designing sample units that can be accurately registered to different types of imagery, confidently located in the field, and classified with objective and repeatable measurements in the field.

Classification systems for remote sensing are often very detailed to meet analytical needs, but classification accuracy can be low. Accuracy increases if similar categories of land cover are combined (i.e., collapsing classification system). A statistically reliable assessment of accuracy can help the analyst decide how to best compromise classification detail without sacrificing the objectives of a specific analysis. This interaction between analysts and their geographic data is extremely important (Goodchild and Gopal, 1989). The composite estimator quickly provides the analyst with an approximate description of the consequences of potential compromises.

ACKNOWLEDGMENTS

I appreciate the assistance and helpful suggestions of Steve Stehman, Oliver Schabenberger, Jim Alegria, Geoff Wood, Russell Congalton, Mike Goodchild, C.Y. Ueng, Hans Schreuder, David Evans, Rudy King, Hans Bodmer, Mohammed Kalkhan, and an anonymous reviewer. Any errors remain my responsibility. This work was supported by the Forest Inventory and Monitoring Program, the National Forest System's Pacific Northwest Region, and the Forest Health Monitoring Program of the USDA Forest Service.

REFERENCES

Agresti, A. Categorical Data Analysis. John Wiley & Sons, New York, 1990.

Arbia, G. The use of GIS in spatial statistical surveys. *Int. Stat. Rev.* 61, pp. 339–359, 1993.

Bauer, M.E., T.E. Burk, A.R. Ek, P.R. Coppin, S.D. Lime, T. A. Walsh, D.K. Walters, W. Befort, and D.F. Heinzen. Satellite inventory of Minnesota forest resources. *Photogramm. Eng. Remote Sens.* 60, pp. 287–298, 1994.

Bishop, Y.M.M., S.E. Fienberg, and P.W. Holland. *Discrete Multivariate Analysis, Theory and Practice.* The MIT Press, Cambridge, MA, 1975.

Brier, S.S. Analysis of contingency tables under cluster sampling. *Biometrika* 67, pp. 591–596, 1980.

Christensen, R. *Linear Models for Multivariate, Time Series, and Spatial Data*. Springer-Verlag, New York. 1991.

Cohen, J.A. Coefficient of agreement for nominal scales. *Edu. Psychol. Meas.* 20, pp. 37–46, 1960.

Cohen, J. Weighted Kappa: Nominal scale agreement with provision for scaled disagreement or partial credit. *Psychol. Bull.* 70, pp. 213–220, 1968.

Congalton, R.G. A review of assessing the accuracy of classifications of remotely sensed data. *Remote Sens. Environ.* 37, pp. 35–46, 1991.

Congalton, R.G. and K. Green. A practical look at the sources of confusion in error matrix generation. *Photogramm. Eng. Remote Sens.* 59, pp. 641–644, 1993.

Costanza, R. and T. Maxwell. Resolution and predictability: An approach to the scaling problem. *Landscape Ecol.* 9, pp. 47–57, 1994.

Czaplewski, R.L. and G.P. Catts. Calibration of remotely sensed proportion or area estimates for misclassification error. *Remote Sens. Environ.* 39, pp. 29–43, 1992.

Czaplewski, R.L. Misclassification bias in areal estimates. *Photogramm. Eng. Remote Sens.* 58, pp. 189–192, 1992a.

Czaplewski, R.L. Accuracy Assessment of Remotely Sensed Classifications with Multi-phase Sampling and the Multivariate Composite Estimator, in Vol. 2, *Proceedings of the 14th International Biometric Conference*. International Biometrics Society, Ruakura Agricultural Centre, Hamilton, New Zealand, 1992b, p. 22.

Czaplewski, R.L. *Variance Approximations for Assessments of Classification Accuracy*. USDA For. Serv. Res. Pap. RM-316, USDA Forest Service, Rocky Mountain Research Station, Fort Collins, CO, 1994.

Fleiss, J.L. *Statistical Methods for Rates and Proportions,* 2nd ed. John Wiley & Sons, Inc. New York, 1981.

Goodchild, M.F. and S. Gopal. *Accuracy of Spatial Databases*. Taylor and Francis, London, 1989.

Graybill, F.A. *Introduction to Matrices with Applications in Statistics*. Wadsworth Publishing Co., Belmont, CA, 1969.

Green, E.J. and W.E. Strawderman. Reducing sample size through the use of a composite estimator: An application to timber volume estimation. *Can. J. For. Res.* 16, pp. 1116–1118, 1986.

Green, E.J., W.E. Strawderman, and T.M. Airola. Assessing classification probabilities for thematic maps. *Photogram. Eng. Remote Sens.* 59, pp. 635–639, 1993.

Gregoire, T.G. and D.K. Walters. Composite vector estimates derived by weighting inversely proportional to variance. *Can. J. For. Res.* 18, pp. 282–284, 1988.

Hoel, P.G. *Introduction to Mathematical Statistics,* 5th ed. John Wiley & Sons, New York, 1984.

Holt, D., T.M.F. Smith, and P.D. Winter. Regression analysis of data from complex surveys. *J. R. Statist. Soc. Ser. A* 143, pp. 303–20, 1980.

Light, R.J. Measures of response agreement for qualitative data: Some generalizations and alternatives. *Psychol. Bull.* 76, pp. 363–377, 1971.

Maybeck, P.S. *Stochastic Models, Estimation, and Control, Volume 1*. Academic Press, New York, 1979.

Molina C., E.A. Measures of Association for Contingency Tables, in *Analysis of Complex Surveys*, C.J. Skinner, D. Holt, and T.M.F. Smith, Eds., John Wiley & Sons, Inc., New York, 1989.

Rao, J.N.K. and D.R. Thomas. Chi-Squared Tests for Contingency Tables, in *Analysis of Complex Surveys,* C.J. Skinner, D. Holt, and T.M.F. Smith, Eds., John Wiley & Sons, Inc., New York, 1989.

Rothman, K.J. *Modern Epidemiology,* Little, Brown and Co., Boston, 1986.

Särndal, C.E., B. Swensson, and J. Wretman. *Model Assisted Survey Sampling,* Springer-Verlag, New York, 1992.

Stehman, S.V. Estimating the Kappa coefficient and its variance under stratified random sampling. *Photogram. Eng. Remote Sens.* 62, pp. 401–402, 1996.

Stehman, S.V. Estimating standard errors of accuracy assessment statistics under cluster sampling. *Remote Sens. Environ.* 60, pp. 258–269, 1997.

Stehman, S.V. Selecting and Interpreting Measures of Thematic Classification Accuracy. *Remote Sens. Environ.* 62, pp. 77–89, 1997.

Stehman, S.V. and R.L. Czaplewski. Design and analysis for thematic map accuracy assessment: Fundamental principles. *Remote Sens. Environ.*, 64(3), pp. 331–334, 1998.

Tenenbein, A.A. Double sampling scheme for estimating from misclassified multinomial data with applications to sampling inspection. *Technometrics* 14, pp. 755–758, 1972.

Van Deusen, P.C. Correcting bias in change estimates from thematic maps. *Remote Sens. Environ.* 50, pp. 67–73, 1994.

Walsh, T.A. and T.E. Burk. Calibration of satellite classifications of land area. *Remote Sens. Environ.* 46, pp. 281–290, 1993.

Williamson, G.D. and M. Haber. Models for three-dimensional contingency tables with completely and partially cross-classified data. *Biometrics* 49, pp. 194–203, 1994.

Section II

Modeling Spatial and Temporal Uncertainty

Chapter 8

Acquisition of Spatial Data by Forest Management Agencies

Michael J.C. Weir

INTRODUCTION

Reliable spatial data are essential for the assessment of forest resources and for planning forest management activities. Because their data requirements are rarely wholly satisfied by topographic and other published maps, many forest management agencies have established facilities for surveying the land under their jurisdiction and for producing maps for internal use.

The acquisition of spatial data by forest management agencies is certainly not a trivial task and accounts for a significant portion of mankind's total effort and investment in resource mapping. For example, each year the forest surveying and mapping authority in Baden-Württemberg, Germany, revises or completely recompiles between 1,500 and 1,800 1:5,000 scale maps covering an area of about 92,000 ha (Holuba, 1995). In Canada, about 30 million ha are inventoried annually by provincial forestry agencies. These agencies maintain more than 37,000 map sheets at scales varying from 1:10,000 to 1:20,000 (Leckie and Gillis, 1995).

Most forest management maps are produced in-house for internal use and technical details of production methods, accuracies, and costs are not widely published. Leckie and Gillis (1995) provide extensive details of current forest mapping practices in Canada. A report by Kennedy et al. (1994) summarizes some forest mapping activities in Europe. Pröbsting (1994) provides a useful review of applications of aerial photography in forest management in Europe.

Over the past 20 years, forest agencies have made increasing use of computer-assisted methods to speed up the tedious tasks of map drafting and area measurement. Although spatial databases support an increasing range of forest management activities, the *raison d'être* for spatial data acquisition by forest management agencies remains the production and revision of maps and the computation of area statistics to support forest inventory (Jordan, 1992).

The aim of this chapter is to examine the techniques currently employed by forest mapping agencies for spatial data acquisition and database maintenance. The presented information is mainly based on data from a limited survey of forestry agencies known to use GIS in production (research organizations were not included). The aim of this survey was to obtain an indication of (i) how forest management agencies acquire spatial data for input to GIS, (ii) the accuracy tolerances (if any) applied, and (iii) policies concerning the updating of their spatial databases. A questionnaire covering these topics was sent to 33 agencies in various parts of the world and 25 replies were received. Most of the replies came from forest agencies in Europe and North America.

The remainder of the chapter is divided into four parts. The first part briefly describes the main categories of spatial data required by forest management agencies and comments on the technology used by the 25 agencies to acquire these data. The second part examines the frequency of forest map revision and database updating. In the third part, specified accuracy tolerances are discussed in the light of the surveying and mapping technology employed. The chapter concludes with some remarks on spatial data accuracy and error modeling in forest management.

SPATIAL DATA ACQUISITION

Information requirements vary from agency to agency depending on forest conditions and management goals. In general, however, forest management maps and spatial databases comprise four main categories of data, namely (i) external (cadastral) boundaries, (ii) topographic details, (iii) "internal" administrative (for example, compartment) boundaries and (iv) information on the forest resources (Table 8.1). Information on spatial data acquisition by the 25 management agencies was compiled for each of these categories of information (Plate 1). It should be noted that the data in the table refer only to the number of instances where a particular technique was reported as being used. The data do not consider the area concerned or the amount of equipment employed. Furthermore, some agencies apply more than one technique to acquire certain categories of spatial data.

Cadastral Information

An exact knowledge of the total area and limits of the forest land is essential for proper management. Particularly in central Europe, large-scale, cadastral maps, supplied by the cadastral survey agency, are the traditional source documents for the compilation of a "basic forest map" (German: *Forstgrundkarte*). In some countries, however, forest management agencies have official cadastral survey responsibilities, such as the determination and demarcation of parcel boundaries in forest areas (Wander, 1981; Holuba, 1995). Even where the forest management agency has no legally based cadastral function, surveys of the legal boundary of the forest may be carried out from time to time. This is particularly the case in developing countries where it is necessary to identify illegal encroachment of agriculture onto the forest land.

Topographic Information

General topographic information is required for planning forest operations, such as road construction and timber harvesting, and to provide a metric framework within which sample plots for forest inventory can be located. Most importantly, topographic features, such as roads, may define (parts of) forest stand boundaries.

Table 8.1. Typical Spatial Information Requirements for Forest Land Management

Administrative boundaries
    cadastral boundaries
    forest administration (e.g., forest district) boundaries
    compartments and subcompartments
    timber concession boundaries

Terrain features
    elevation
    slope
    aspect
    drainage

Infrastructure
    roads, tracks, etc.
    power lines, pipelines, etc.
    buildings and other structures

Forest stand characteristics
    species composition
    age
    yield class
    density

Location of management activities
    realized and planned treatment (thinning, harvesting, etc.)
    land use zoning (e.g., nature protection)
    fire control facilities
    damage control (e.g., insects, wind)

Topographic information therefore forms an integral part of any (graphical or digital) forest database rather than simply a background against which details, such as forest types, are mapped. The survey of 25 forest agencies showed that the primary source of this information is maps and digital data obtained from topographic mapping agencies. However, additional surveys by forest agencies themselves are generally required to map details which are not shown on general purpose topographic maps and to locate "new" features such as recently constructed forest roads. Land surveying, GPS, and a range of photogrammetric methods are all employed for this purpose. Some agencies report using relatively approximate methods of graphical transfer from aerial photographs for completing topographic details. In such cases, the spatial accuracy is likely to be lower than that achieved by the national mapping agency, leading to inconsistencies in quality.

Permanent Boundaries Established by Forest Agencies

Many forest management agencies have jurisdiction over large areas of land. For this reason, they may establish and maintain a "quasi cadastral" survey in the form of compartments or timber concession areas. The demarcation, survey, and mapping of these boundaries can be a considerable task. For example, one forest management agency reported employing 36 land surveyors who, in a typical year, carry out relocation surveys of some 30,000 property boundary pillars, the survey of 3,000 km of compartment boundary, and the survey of subcompartment boundaries over about 45,000 ha of forest land.

The survey showed that a wide range of surveying and mapping techniques is employed to acquire data on permanent "internal" forest boundaries. Unlike property boundaries which, for reasons of accuracy, are primarily surveyed in the field, permanent "internal" forest boundaries need not be surveyed to cadastral mapping standards. Photogrammetric techniques are therefore feasible and, according to the survey, predominate over field surveys.

Forest Information

Information on stand boundaries and other specialized forest information, such as the location of planned thinning or harvesting operations, is rarely available from existing sources. Acquisition of this category of information is therefore one of the main tasks of any forest surveying and mapping agency. According to the survey, just over half of the agencies employ simple methods of graphical transfer from aerial photographs for this purpose. The operating costs of these "simple" graphical methods are not significantly lower than those associated with the use of rigorous photogrammetric methods (Weir, 1981). It is therefore surprising to find this labor-intensive approach still widely used by forest management agencies in developed countries. There are, however, indications that these graphical methods will be replaced by digital, "softcopy," photogrammetry in the foreseeable future.

Only 2 of the 25 agencies reported using satellite imagery for mapping forest details. The spatial resolution of the current generation of earth observation satellites is apparently inadequate for mapping at the scale and level of detail required by many forest management agencies in Europe and North America. For example, Delaney and Archibald (1993) describe an experiment on map revision in which the use of SPOT panchromatic imagery gave an accuracy of ±34 m and a 41% completeness of change detection compared with ±5 m and 91% for photogrammetric revision using an analytical plotter and 1:50,000 aerial photography. The costs of the two approaches were almost the same. It remains to be seen whether aerial photography will be able to maintain this advantage in the face of competition from imagery acquired by the coming generation of high resolution satellites (Aplin et al., 1997). Particularly in the developing countries, where funds for the acquisition of aerial photography are limited, high resolution satellite imagery may offer a timely, and more affordable, source of high quality spatial data for forest monitoring and resource management.

FREQUENCY OF REVISION AND DATABASE UPDATING

Unlike a soil or geological map which, if carefully produced, may serve as a useful document for many decades, a forest management map must be updated at regular intervals. In the intensively managed forests of Western Europe, where management plans are traditionally prepared every 10 years, management maps are revised on a corresponding 10-year cycle.

Experience with database updating by forest management agencies is still limited. At the time of the survey (mid-1994), only two systems had been in operation for more than 10 years. Eight agencies had being using their system for less than five years and some agen-

Table 8.2. Frequency of Database Updating by 16 Forest Management Agencies

	Cadastral Boundaries	Topographic Features	Internal Boundaries	Forest Boundaries
Continuous	7	9	9	5
Several times per year	–	1	–	1
Annually	3	3	3	4
Every 2–5 years	2	1	1	1
Every 10 years	2	1	3	5

cies had not yet done any updating. Table 8.2 gives details of the frequency of updating employed by 16 of the surveyed agencies who have been using GIS long enough to have gained some experience of database maintenance. The table indicates a tendency toward continuous or annual updating. An example of this trend is provided by Corrie et al. (1994) who published details of the database maintenance procedures employed by a timber company for 2.4 million ha of productive forest in British Columbia. The database, which includes 250 base and thematic layers, is continuously updated, with all changes distributed to local offices on a nightly basis.

Interestingly, three agencies indicated that forest stand information is updated by means of a complete remapping rather than by considering only the actual changes. Although this would appear to negate the advantages of a digital database, further research is needed to clarify this issue. One agency, which has been employing GIS since 1977, updates their database on a 2-year cycle but makes a complete overhaul every 10 years.

ACCURACY STANDARDS

The operating costs of a large GIS to support forest management "pale in contrast" to the size of the investment decisions supported by the spatial database (Jordan and Erdle, 1989, p. 294). Nevertheless, the size of these decisions is such that forest management agencies must consider the impact of spatial data quality (Keefer, 1994; Prisley, 1994). This applies particularly to the area data used in calculating timber volume, and thus resource value. Compared to the efforts taken to reduce the sampling errors associated with forest inventory, however, only scant attention is paid to the question of area accuracy (Prisley and Smith, 1991). As Gemmell et al. (1991, p. 221) note, "most existing forest inventories do not have estimates of the accuracies of their polygon labels and boundaries."

The survey of 25 forest management agencies revealed a wide range of accuracy tolerances, ranging from a 50 m "worst estimate" for forest stand boundaries down to submeter tolerances for cadastral boundaries. As indicated in Plate 1, however, the latter category is often provided by other agencies to predetermined quality standards. The other three categories of information are mainly acquired by forest management agencies themselves, usually to a less strict tolerance than that set for cadastral information. Only four agencies reported applying the same tolerance to all categories of data. Leckie and Gillis (1995, p. 81) report that most Canadian forest inventory maps are considered to have accuracies "in the 10 to 25 m range." In Europe, accuracy tolerances for data acquired by forest management agencies are generally in the range 2–10 m rms error relative to nearby fixed detail, although some agencies reported that they relax the tolerances for—relatively uncertain—forest stand boundaries.

Although almost all of the agencies reported that they define accuracy tolerances for the acquisition of spatial data, almost half (12) indicated that they do not apply any quality control measures to check if these tolerances are actually being met. Significantly, 8 of these 12 agencies employ simple graphical methods for mapping forest boundaries from aerial photographs. Tests with a variety of instruments, photo-scales and operators (Weir, 1981) have shown that the accuracy of these methods (under optimal conditions, 0.8 to 1.3 mm rms plotting error) is highly sensitive to terrain conditions, base map quality, and operator skill. The inconsistent performance of simple photogrammetric instruments therefore makes them inappropriate for mapping forest stand boundaries at a typical scale of 1:10,000 if a (again, typical) accuracy tolerance of ±5 m rms error has been specified.

DISCUSSION

It has been suggested that "the environmental planning, management and design professions rely extensively on data collected and published by specialized data collection agencies" and that "it is only rare that new data is actually collected by original field measurement" (Sinton, 1978, p. 2). In this respect, forestry differs from most other fields of planning and resource management. Because they are directly involved in primary data acquisition for the establishment and maintenance of their own graphical and/or digital spatial databases, forest management agencies are in the fortunate position of being able to exert a considerable degree of control over spatial data quality.

The previous section has shown that there is a need for forest management agencies to define standards for spatial data acquisition more clearly and, in particular, to apply appropriate quality control measures. In recent years, however, GIS researchers have laid much of the groundwork needed to handle and predict the propagation of spatial error in natural resource databases. By chaining back through the various GIS operations, the accuracy of primary data acquisition needed to achieve the desired quality of forest information products can be determined. Nevertheless, this will not be an easy task. Spatial databases maintained by forest management agencies typically contain data varying from precisely determined cadastral boundaries to uncertain regions such as areas of different susceptibility to windthrow. Frequently, different methods are used to acquire these data so that, in the longer term, spatial databases will evolve to contain data of widely varying age and lineage. Common standards such as the Spatial Data Transfer Standard (SDTS) make little or no allowance for variations in quality from one object to another (Buttenfield and Beard, 1994). Indeed, it is questionable whether any one method of assessing spatial data quality can be successfully applied to all categories of data in a typical forestry database which is used to support a wide range of decision-making.

As this chapter has shown, many forestry agencies work to, or at least aim at, "near professional" standards of surveying and mapping. High quality data are, however, relatively expensive. In the case of forest inventory, it has been suggested that "allowable sampling errors are frequently established by the inventory specialist" (Lund, 1990, p. 6), rather than by the decision-makers who use the data. Consequently, spatial data acquisition (not, as such, a profitable activity) may be a candidate for budget cuts unless a sensible balance between forest resource value and the methods and costs of surveying and mapping can be achieved.

ACKNOWLEDGMENTS

This chapter is based on information kindly supplied by forestry agencies in many parts of the world. An anonymous referee and my colleagues, Prof. A. De Gier and Dr. Y. Hussin, have provided useful suggestions for improving an earlier version of the manuscript.

REFERENCES

Aplin, P., P.M. Atkinson, and P.J. Curran. Fine resolution satellite sensors for the next decade. *Int. J. Remote Sensing,* 18(18), pp. 3873–3881, 1997.

Buttenfield, P.B. and M.K Beard. Graphical and Geographical Components of Data Quality, in *Visualization in Geographical Information Systems,* H. Hearnshaw and D. Unwin, Eds., Wiley/Belhaven, London, 1994.

Corrie, G., W. Reedijk, and K. Lohia. A GIS to Improve the Competitive Advantage of TimberWest Forest Limited, in *Proceedings GIS '94,* Vancouver, 1994, pp. 83–88.

Delany, B.B. and J.H. Archibald. Towards an Update Methodology for Integrated Databases (Digital Base Mapping and Vegetation Inventories), in *Proceedings of the International Forum on Airborne Multispectral Scanning for Forestry and Mapping,* Petawawa National Forestry Institute, Chalk River, 1993, pp. 111–116.

Gemmell, F., D.G. Goodenough, K. Fung, and C. Kushigbor. Resource Spatial and Attribute Information Extraction from Remotely Sensed Data, in *Proceedings Spatial Data 2000,* Oxford, 1991, pp. 221–231.

Holuba, K.H. Forstvermessung und Forstkartographie in Baden-Württemberg. *Allgemeine Forst Zeitschrift* 22/95, 1995, pp. 1198–1200.

Jordan, G.A. and T.A. Erdle. Forest management and GIS: What have we learned in New Brunswick? *CISM Journal,* 43(3), pp. 287–295, 1989.

Jordan, G. GIS and Forest Management—Towards the Next Century, in *Proceedings GIS '92,* Vancouver, 1992.

Keefer, B.J. Impact of Spatial Accuracy on Business Decisions: A Forest Management Perspective, in *Proceedings International Symposium on the Spatial Accuracy of Natural Resource Data Bases,* Williamsburg, 1994, pp. 18–23.

Kennedy, P.J., R. Päivinen, and L. Roihuvuo, Eds. *Proceedings International Workshop on Designing a System of Nomenclature for European Forest Mapping,* European Commission, Luxemburg, 1994.

Leckie, D.G. and M.D. Gillis. Forest inventory in Canada with emphasis on map production. *For. Chron.,* 71(1), pp. 74–88, 1995.

Lund, H.G. The Platonic Verses and Inventory Objectives, in *State-of-the-art Methodology of Forest Inventory,* U.S. Department of Agriculture, Forest Service Pacific Northwest Research Station, Technical Report PNW-GTR-263, 1990.

Prisley, S.P and J.L. Smith. The Effect of Spatial Data Variability on Decisions Reached in a GIS Environment, in *Proceedings GIS '91,* Vancouver, 1991.

Prisley, S.P. Why Natural Resource Information Managers Must Be Concerned about Spatial Accuracy, in *Proceedings International Symposium on the Spatial Accuracy of Natural Resource Data Bases,* Williamsburg, 1994, pp. 24–34.

Pröbsting, T. Einsatz von Luftbildern in der Forstwirtschaft in Europa—Ein Überblick in *Photogrammetrie & Forst,* T. Pröbsting, Ed., Albert-Ludwigs-Universität, Freiburg, 1994, pp. 277–296.

Sinton, D. The Inherent Structure of Information as a Constraint to Analysis: Mapped Thematic Data as a Case Study, in *Harvard Papers on Geographic Information Systems,* Vol. 6, G. Dutton, Ed., Addison-Wesley, Reading, MA, 1978.

Wander, R. Die Forstvermessung in Landesteil Baden, *Allgemeine Vermessungsnachrichten,* 8/9, 1981, pp. 329–338.

Weir, M.J.C. Simple Plotting Instruments for Resource Mapping, in *Proceedings of the International Conference on Matching Remote Sensing Technologies and their Applications,* London, 1981, pp. 223–232.

Chapter 9

Choosing Between Abrupt and Gradual Spatial Variation?

Gerard B.M. Heuvelink and Johan A. Huisman

INTRODUCTION

Traditional soil surveying is based on the presumption that soil behaves uniformly within soil mapping units and changes fairly abruptly at the boundaries between them (Voltz and Webster, 1990; Webster and Oliver, 1990; Burrough, 1993). However, the practical validity of this conventional representation of soil variability has repeatedly been questioned (e.g., Webster and De La Cuanalo, 1975; Nortcliff, 1978; Campbell et al., 1989; Nettleton et al., 1991). Major drawbacks of the conventional model of soil spatial variation are that it cannot represent gradual boundaries and that it ignores spatial autocorrelation within and across mapping units. As an alternative to the conventional model, some 20 years ago the geostatistical approach to modeling spatial variation was introduced to soil science. Burgess and Webster (1980) were among the first to apply kriging, an exponent of this theory, to soil surveying. The geostatistical approach has proved very suitable for modeling soil spatial variation, and kriging is now routinely being applied to construct maps of soil properties from point observations (Webster, 1994). Recently, however, it is more and more being realized that the outright abandoning of the conventional approach to handling soil spatial variation may not be that sensible after all. Ordinary kriging definitely has disadvantages as well, such as its inability to deal with sharp boundaries.

In order to bridge the gap between the conventional and geostatistical representation of soil spatial variation, several models have been proposed to deal with a situation in which there is discrete (abrupt) as well as continuous (gradual) spatial variation in the same area (Stein et al., 1988; Voltz and Webster, 1990; Heuvelink and Bierkens, 1992; Rogowski and Wolf, 1994; Goovaerts and Journel, 1995; Heuvelink, 1996). In this chapter we examine one such a model, known as the *Mixed Model of Spatial Variation* (MMSV). In Heuvelink (1996) it was anticipated that the MMSV should work well on the whole range of spatial variation, from purely discrete to purely continuous variability. We analyze the anticipated flexibility of the MMSV using nine simulated 'realities.' But before describing the exact procedure of the simulation study, we will first briefly review the three models of spatial variation that are used in this study.

THREE MODELS OF SPATIAL VARIATION

The *Discrete Model of Spatial Variation* (DMSV) first divides the geographical domain D into K separate units D_k (k=1,...,K). It then makes the following assumptions about the behavior of a spatially distributed attribute $Z(x)$:

- $Z(x)=\mu_k+\varepsilon(x)$ for all $x \in D_k$
- $\varepsilon(x)$ has zero mean and is spatially uncorrelated
- $Var(\varepsilon(x))=C_0$ for all $x \in D$

Thus the DMSV assumes that the random function $Z(x)$ is the sum of a unit-dependent mean μ_k and a residual noise $\varepsilon(x)$. The DMSV will usually be adopted when the units D_k are available in the form of a polygon map, such as a soil map, a land use map or a geological map, and when the within-unit variability is expected to be small in comparison with the between-unit variability. In other words, the DMSV represents the conventional model of soil spatial variation and is appropriate when major jumps in the attribute $Z(x)$ take place at the boundaries of the mapping units. For instance, many physical and chemical soil properties depend on soil type and so the DMSV may be adopted using the soil map to partition the area (Beckett and Burrough, 1971; Van Kuilenburg et al., 1982).

In its simplest form, the *Continuous Model of Spatial Variation* (CMSV) makes the following assumptions ($E[\cdot]$ stands for mathematical expectation):

- $E[Z(x)]=\mu$ for all $x \in D$
- $Cov(Z(x),Z(x+h))=C_Z(|h|)$ for all $x,x+h \in D$

Thus the CMSV assumes that $Z(x)$ is second-order stationary, meaning that it has a constant mean and that its spatial autocovariance is a function only of the distance between the locations. The CMSV embodies the geostatistical model of spatial variation. Note that in the geostatistical literature it is customary to use the variogram $\gamma_Z(x)$ to characterize the spatial autocorrelation of $Z(x)$. It is related to the autocovariance function by the identity $\gamma_Z(|h|)=C_Z(0)-C_Z(|h|)$.

The assumptions underlying the *Mixed Model of Spatial Variation* (MMSV) are a combination of those underlying the DMSV and CMSV:

- $Z(x)=\mu_k+\varepsilon(x)$ for all $x \in D_k$
- $\varepsilon(x)$ has zero mean
- $Cov(\varepsilon(x),\varepsilon(x+h))=C_\varepsilon(|h|)$ for all $x,x+h \in D$

Second-order stationarity is thus imposed on $\varepsilon(x)$ instead of on $Z(x)$. The MMSV is more general than the DMSV and CMSV, and in fact it contains both models. However, the DMSV and CMSV are included here as separate models because they are very often used in practice.

APPLICATION TO NINE SIMULATED REALITIES

In order to study the suitability of the three models of spatial variation for mapping under various conditions, we created nine different 'realities.' We did this by adding maps generated using unconditional sequential Gaussian simulation (Deutsch and Journel, 1992) to an artificially constructed 'soil map.' The nine maps are given in Plate 2. The A-maps in Plate 2 (top row) are strongly dominated by the discrete soil map, the B-maps (middle row) to a much lesser extent, and the C-maps (bottom row) are not influenced at all by the discrete soil map. The degree of spatial autocorrelation in the added residual decreases from the 1-maps (left column) to the 3-maps (right column).

Mapping the Soil Property from 200 Observations

From each of the nine simulated realities, data sets were created by collecting observations at 200 randomly selected locations. From these observations the soil property was mapped using the three MSVs. This resulted in 27 maps of predictions and 27 maps of prediction error standard deviations. Note that the DMSV and MMSV need as input not only the 200 observations, but also a polygon map delineating the soil mapping units.

Mapping using the DMSV is done simply by calculating the mean of the observations for all mapping units separately, and using the unit mean as a prediction for all points lying in the same unit. Mapping with the CMSV is done by ordinary kriging, whereby a variogram is used derived from the 200 observations. Mapping is somewhat more complicated in case of the MMSV (Heuvelink, 1996). First a variogram is derived from the 200 residuals, which have been calculated by subtracting the unit means from the observations (Kitanidis, 1994). Next the attribute is mapped using universal kriging (Cressie, 1991).

In Plate 3 the mapping results are given for a selection of three out of the nine simulated realities. Note that these are the prediction maps and that the corresponding prediction error maps are not given here. The DMSV maps necessarily follow the delineations of the soil map, which is quite all right for reality A3 but less so for reality B2 and definitely inappropriate for reality C1. Conversely, the CMSV is the appropriate model for mapping reality C1, but it is not appropriate for mapping B2 and even less so for mapping A3. The most important observation from Plate 3 is that the MMSV is indeed capable of an adequate mapping in all three cases. It is interesting to observe that the MMSV mimics the DMSV in case of a 'discrete reality' and that it mimics the CMSV in case of a 'continuous' reality.

Validation

In order to evaluate the three prediction methods quantitatively, the mean error (ME), root mean square error (RMSE), and standardized root mean square error (SRMSE) were computed for all nine realities. These statistics were computed from all remaining points in the map. The results are given in Figure 9.1.

The mean error is in all cases quite small. This is not surprising, because unbiasedness conditions are included in all three mapping procedures. Differences between the three mapping procedures are also negligible.

The SRMSE values are on average somewhat larger than one, particularly when the CMSV is applied to realities where the soil map has a strong influence. This may be caused by forcing the CMSV upon a nonstationary MMSV or DMSV reality, and perhaps also because the kriging variance does not include the uncertainty in estimating the variogram (Christensen, 1991).

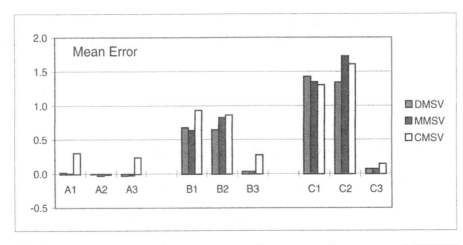

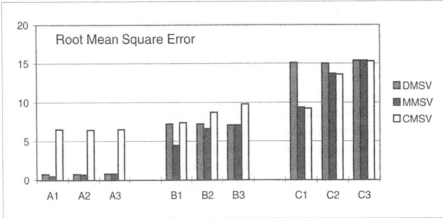

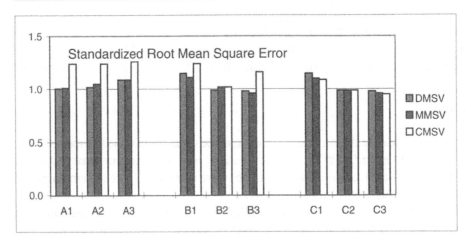

Figure 9.1. Validation results for mapping the nine simulated realities under the three models of spatial variation.

Most interesting are the RMSE results. In this case we do see meaningful differences between the three mapping procedures. The results demonstrate that the CMSV is a very poor model for the A-realities, whereas the DMSV is poor for the C-realities. An exception is the pure nugget map C3, where, as could be foreseen, all mapping procedures are equally good (or bad). The results also confirm that the MMSV is superior for the B-realities. Note that comparison of RMSE values between realities is difficult here because these are affected by differences in levels of spatial autocorrelation.

DISCUSSION AND CONCLUSIONS

The application to simulated 'realities' shows that the MMSV interpolates well on the entire range of spatial variation. In all cases the MMSV is at least as good as the DMSV and the CMSV, and it is superior in situations where there is both abrupt and gradual spatial variation.

The simulation exercise shows that the DMSV will outperform the CMSV when abrupt spatial variation dominates over gradual spatial variation. The CMSV will strongly outperform the DMSV in situations where gradual spatial variation is dominant. This means that when a choice between these two models is to be made (and these are the two models that are most often used in practice), then it must be taken with care. This may seem obvious, but in practice the choice of model is often dominated by irrelevant factors, such as background and experience of the user. And even when care is taken, it may not always be easy to decide beforehand whether abrupt or gradual spatial variation prevails. Therefore the flexibility of the MMSV demonstrated here is of clear importance. It implies that by adopting the MMSV one can protect oneself against using a poor model. In other words, one does not need to choose beforehand between abrupt or gradual spatial variation but instead can leave the decision to the MMSV.

In formulating the DMSV we have assumed that the within-unit variance C_0 is the same for all units. Alternatively, this assumption may be weakened by allowing the variance to vary between units. This should be recommended when the within-unit variability is not the same for all units. The advantage is a more general model that can handle these differences, but the disadvantage is that per unit variances are based on smaller samples and are thus less unreliable.

The MMSV is designed for situations where there is abrupt and gradual spatial variation. Some indication of whether this is the case can be obtained from the intraclass correlation (Webster and Oliver, 1990), but more informative is the comparison of the variograms of the original attribute and its residual. In a situation of mixed spatial variation, the sill of the variogram of the residual will be substantially smaller than that of the original attribute and it will still be showing spatial correlation (i.e., it will not be a pure nugget variogram). Since the variograms of the original attribute and its residual can easily be computed, one can thus quickly decide whether the MMSV is superior to the DMSV and CMSV for a given situation.

Another approach to handling the presence of abrupt and gradual spatial variation is known as stratified kriging (Stein et al., 1988; Voltz and Webster, 1990). In this approach the CMSV is adopted for each mapping unit separately. The main difference with the MMSV is that stratified kriging excludes the presence of spatial autocorrelation across mapping unit boundaries. Therefore, it is prone to creating boundaries when they are not really there.

As demonstrated, the MMSV does not suffer from this problem because it mimics the CMSV under such circumstances. We consider it a major advantage of the MMSV that, although it is meant for situations in which gradual and abrupt spatial variation are both present, it will also perform well when spatial variation is exclusively gradual or exclusively abrupt.

A problem not addressed in this chapter is the positional accuracy of the delineations in the soil map. Both the DMSV and the MMSV assume that the boundaries of the soil map are exactly known and error-free. Clearly, in practice there will be uncertainty about the boundaries' location. Recently, much progress has been made in handling positional uncertainty in polygon maps (Edwards and Lowell, 1996; Kiiveri, 1997; Boucneau et al., 1998). It would be interesting to investigate whether the methods described in these studies and the properties of the MMSV could be unified into one model of spatial variation and thus improve our ability to model the complex realities that we so often encounter in the earth and environmental sciences.

REFERENCES

Beckett, P.H.T. and P.A. Burrough. The relation between cost and utility in soil survey. IV: Comparison of the utilities of soil maps produced by different survey procedures, and to different scales, *J. Soil Sci.* 22, pp. 466–480, 1971.

Boucneau, G., M. Van Meirvenne, O. Thas, and G. Hofman. Integrating properties of soil map delineations into ordinary kriging, *Eur. J. Soil Sci.* 49, pp. 213–229, 1998.

Burgess, T.M. and R. Webster. Optimal interpolation and isarithmic mapping of soil properties. I. The semi-variogram and punctual kriging, *J. Soil Sci.* 31, pp. 315–331, 1980.

Burrough, P.A. Soil variability: A late 20th century view, *Soils Fert.* 56, pp. 529–562, 1993.

Campbell, D.J., D.G. Kinninburgh, and P.H.T. Beckett. The soil solution chemistry of some Oxfordshire soils: Temporal and spatial variability, *J. Soil Sci.* 40, pp. 321–339, 1989.

Christensen, R. *Linear Models for Multivariate, Time Series, and Spatial Data*, Springer, New York, 1991.

Cressie, N. *Statistics for Spatial Data*, Springer, New York, 1991.

Deutsch, C.V. and A.G. Journel. *GSLIB: Geostatistical Software Library and User's Guide*, Oxford University Press, New York, 1992.

Edwards, G. and K.E. Lowell. Modeling uncertainty in photointerpreted boundaries, *Photogrammetric Eng. Remote Sensing*, 62, pp. 337–391, 1996.

Goovaerts, P. and A.G. Journel. Integrating soil map information in modelling the spatial variation in continuous soil properties, *Eur. J. Soil Sci.* 46, pp. 397–414, 1995.

Heuvelink, G.B.M. Identification of field attribute error under different models of spatial variation, *Int. J. GIS* 10, pp. 921–935, 1996.

Heuvelink, G.B.M. and M.F.P. Bierkens. Combining soil maps with interpolations from point observations to predict quantitative soil properties, *Geoderma.* 55, pp. 1–15, 1992.

Kiiveri, H.T. Assessing, representing and transmitting positional uncertainty in maps, *Int. J. GIS* 11, pp. 33–52, 1997.

Kitanidis, P.K. Generalized covariance functions in estimation, *Math. Geol.* 25, pp. 525–540, 1994.

Nettleton, W.D., B.R. Brasher, and G. Borst. The taxadjunct problem, *Soil Sci. Soc. Am. J.* 55, pp. 421–427, 1991.

Nortcliff, S. Soil variability and reconnaissance soil mapping: A statistical study in Norfolk, *J. Soil Sci.* 29, pp. 403–418, 1978.

Rogowski, A.S. and J.K. Wolf. Incorporating variability into soil map unit delineations, *Soil Sci. Soc. Am. J.* 58, pp. 403–418, 1994.

Stein, A., M. Hoogerwerf, and J. Bouma. Use of soil-map delineations to improve (co-) kriging of point data on moisture deficits, *Geoderma* 43, pp. 163–177, 1988.

Van Kuilenburg, J., J.J. De Gruijter, B.A. Marsman, and J. Bouma. Accuracy of spatial interpolation between point data on soil moisture supply capacity, compared with estimates from mapping units, *Geoderma* 27, pp. 311–325, 1982.

Voltz, M. and R. Webster. A comparison of kriging, cubic splines and classification for predicting soil properties from sample information, *J. Soil Sci.* 31, pp. 505–524, 1990.

Webster, R. The development of pedometrics, *Geoderma.* 62, pp. 1–15, 1994.

Webster, R. and H.E. De La Cuanalo. Soil transect correlograms of North Oxfordshire and their interpretation, *J. Soil Sci.* 26, pp. 176–194, 1975.

Webster, R. and M.A. Oliver. *Statistical Methods in Soil and Land Resource Survey,* Oxford University Press, Oxford, 1990.

Chapter 10

Modeling the Spatial Distribution of Ten Tree Species in Pennsylvania

Rachel Riemann Hershey

INTRODUCTION

Current, accessible information on the distribution of tree species is an important element in both the research and management of forest ecosystems. Ground inventory remains a unique and necessary source for much of this detailed information we require about the forest resource. In addition, there is a need for continuous spatial output, for maps depicting where and how those forest attributes are distributed across the landscape, to address the spatial questions of forest management and regional planning. Extensive and periodic data for individual species are available from the Forest Inventory and Analysis (FIA) inventory, but only at sample plot locations. Creating a map of this information requires a model of the resource so that we can interpolate information between known locations using the information available. Basic spatial summaries of the data, such as county averages, have frequently been used. However, such aggregation can mask real spatial patterns because of the effect of the choice of area size and shape in the aggregation (Monmonier, 1991). In addition, such averages often do not provide a picture of the spatial variability that actually exists in the landscape, which can be an important factor in both planning decisions and research assumptions. Individual species were targeted specifically in this study because forest composition is often poorly characterized by the discrete categories imposed by forest-cover-type divisions. Inherent in each forest-type category is an entire continuum (usually multidimensional) of different species and their relative importance.

Geostatistical techniques were used to examine and incorporate the data's inherent spatial structure into local area estimates. Geostatistics offers a set of tools and techniques that can be applied when trying to understand the spatial characteristics of a phenomenon, or when trying to estimate it in a spatial context. Many useful references exist, both within the earth sciences where geostatistics have been widely used and developed (Isaaks and Srivastava, 1989; Deutsch and Journel, 1992; Srivastava, 1994; Wackernagel, 1995; Goovaerts, 1997), and increasingly in ecology and forestry (Samra et al., 1989; Rossi et al., 1992; de Fouquet and Mandallaz, 1993; Liebhold et al., 1993; Mowrer, 1994). In a previous study (Riemann Hershey et al., 1997), we compared the geostatistical techniques available to interpolate FIA sample data to create a 'map' of tree species distribution. Each of these estimation techniques honors and maintains different aspects of the original data. The specific goals of the interpolation task at hand and the characteristics of the data and the phenomena will determine the priorities, but in general the more characteristics of the sample data that are preserved in the estimated data set, the more potential the data set(s) has for general use. In addition to providing an estimate of species occurrence and relative importance, several other features were considered necessary to make such a data set/map really useful—the ability to produce a measure of the uncertainty associated with the modeled estimate and the ability to maintain and/or provide some indication of the variability that really exists. Other important factors were the ability to provide some assessment of model performance, and the ability to reduce the sensitivity of the model to variables we knew little about. Indicator kriging (IK) and sequential Gaussian conditional simulation (sgCS) were found to be the best interpolation tools for this task, and to provide the most complete set of useful output information, given our interests in:

- providing an estimate of species occurrence,
- providing an estimate of species relative 'importance' or density,
- providing a measure of uncertainty associated with the estimates,
- maintaining some indication of local variability,
- maintaining the univariate and spatial characteristics of the original sample data (e.g., interested in maintaining the extremes), and,
- handling sample data with highly skewed distributions.

There is uncertainty associated with any measurement or estimate. Knowing that uncertainty provides us with information as to how much we do not know. It affects how much weight we give to different sources of information in both our decisions and our analyses. And in data sets modeled to broader spatial scales, where the spatial resolution/detail being reported is necessarily a summary of local conditions, the uncertainty provides an indication of the magnitude of that local variation. Knowing how much uncertainty exists helps the user identify whether that uncertainty is acceptable for a specific task and how the data can be used. In addition, knowing how much uncertainty exists helps the producer identify areas in which additional sampling, or additional related data, would most improve the estimates.

Understanding the variability that actually exists in a landscape is also important information for both management and analysis. Making explicit this variability, as an additional characteristic of the output, is useful additional information for the user. A highly variable class may be treated very differently from a more homogeneous one, and it may have different implications for the types of management that are appropriate under those conditions. Tree species in Pennsylvania do exhibit a high level of local variation as a result of small-scale topographic and other environmental factors and land-use histories. At the intensity of sampling present in the FIA sample data, much of this local variation cannot be modeled and effectively predicted by spatial information alone, and instead appears as variation that is unexplained by neighboring plots. However, such local variability is an important characteristic of the distribution of a species. Thus, we did not want this local variability to become hidden behind a regional average of the resource, but to remain as apparent and accessible to the user as possible in the final estimated data set(s).

The previous study offered promising results for using geostatistical techniques for estimating the distribution of sugar maple from FIA data. Thus, the same geostatistical methods were applied in this study to nine additional species: red oak, white oak, chestnut oak, black oak, hemlock, red maple, beech, white pine, and yellow birch. This list includes 8 of the top 10 most abundant species in Pennsylvania by volume, and 2 species (yellow birch and white pine) that are much less common (Alerich, 1993) to investigate the effect on the accuracy of estimating a species occurring only relatively rarely.

Tree species distribution is affected by many factors, including both environmental conditions and direct human influence through harvesting and other land use histories. As a result of being differentially affected by all of these factors, each species exhibits different patterns and scales of spatial distribution. Some of these factors occur at scales much smaller than the sampling intensity of the FIA data, and some occur over larger areas, representing broad-scale variation in the species distribution. In the previous study, it was found that a substantial amount of variation in sugar maple distribution (97%) was resolved at the sampling scale used for the FIA plots. This spatial dependence could, therefore, be modeled and used to support estimates of species occurrence and relative "importance" (% ba/acre). The goal of this study is to examine to what extent this is true for the other species. More specifically, the objectives of this study are to determine:

1. if spatial dependence of tree species (% ba/acre) is exhibited at this intensity/scale of sampling,
2. how tree species differ from one another in terms of spatial dependence and distribution, and how that affects our ability to estimate them,
3. what the resulting spatial distribution for each species is, and how that compares to our current understanding,
4. how the resulting estimated data sets and uncertainty information can be used, and
5. what possibilities exist for improving the estimates.

DATA

The sample data were collected by the USFS Northeastern Research Forest Experiment Station's Forest Inventory and Analysis (FIA) unit. Basal area—cross-sectional area at breast height (4-1/2 feet)—is calculated for all live trees 1.0 inches DBH or larger on the plot (Hansen et al., 1992). The data were calculated for individual tree species, by basal area (ba) per acre as a proportion of the total basal area (% ba/acre). The data were accessed from individual tree records in the USFS Eastwide database and summarized as % ba/acre for each species by plot. In Pennsylvania, there were 5,100 plots. Nonforested plots and those with total ba/acre equal to zero as a result of missing data were removed—2,905 plots remained and were used in the analysis. The distinction between forest and nonforest areas was found to occur at a finer scale than was resolved by the FIA plots, and was considered to be better derived from another source.

One feature of a well-designed sampling scheme is that it is sensitive to and can report, with an acceptable level of error, the characteristics of the phenomena of interest. In this ideal situation, the characteristics of the sample data represent reasonably well those characteristics of the phenomenon itself. The FIA data were considered to be an unbiased sample of forested areas in this analysis.

METHODS

Each species was examined independently. As in the previous study, the data for each species were organized, summarized, and explored using univariate statistics, measures of spatial dependence (variogram, covariance, and correlogram), and spatial distribution of local statistics across the state. The resulting information was critical not only in determining what interpolation methods were most suitable and for checking the sample data for errors, but also for understanding the characteristics of the sample data and thus the phenomena being investigated. All species were similar in many of their basic characteristics to the previously investigated species, sugar maple. Each species exhibited positively skewed distributions, with more than 50% of the plots containing less than 1% ba/acre in every species except red maple and red oak.

In this study the data were transformed to a normal distribution for use in conjunction with the conditional simulation. The normal-score transform used was a nonlinear transform, but it provided a 1-to-1 mapping of values to allow for an accurate back transformation to % ba/acre values (Deutsch and Journel, 1992). A variogram was calculated for both the raw sample data and for the normal-scored data, using a lag distance of 500 m and no directional component (anisotropy). Although there may be some directional differences in local areas, such as in the NE-SW direction in the ridge and valley section, or in the E-W direction of the Allegheny Plateau, the mixture and complexity of topography in Pennsylvania are such that when the state is treated as a single region, little anisotropy is exhibited in the sample variogram. Such regional differences in spatial pattern provide additional evidence for dividing the area of interest into smaller regions for modeling and simulation, but this more time-consuming approach was not used in this study. In every instance, the sample variogram for the normal-scored data exhibited considerably more spatial dependence and structure than for the raw data (Figure 10.1), revealing a spatial structure that had been hidden by the strong positive skew of the data. As sgCS uses normal-scored data, it was the model fitted to the normal-scored sample variogram that was used in the conditional simulation. An indicator variogram also was calculated and modeled for use in the indicator kriging. To assess how areas of 'local' variability in the sample data changed across the state, the mean and standard deviation were calculated for each of the 23,400 3,000- × 3,000-m cells, using a 15- × 15-km area as the window defining the size of the 'local' area.

The two geostatistical interpolation methods used were indicator kriging (IK) and sequential Gaussian conditional simulation (sgCS). As found in a previous study, these two techniques proved to be the most appropriate and effective tools for interpolation, given the objectives and the characteristics of the data and the phenomena of interest. The estimate of the relative amount of sugar maple at that location—i.e., whether the species represented a minor, moderate, or a major component of the total ba/acre on the plot at that location— was provided by sgCS. Sequential Gaussian conditional simulation determines multiple estimates for each cell. All are equally probable, and yet alternative realizations of the data determined from multiple simulation runs. From this set of estimates, an entire distribution function can be built for each cell, representing the range of possible values. A summary statistic such as the mean or median of this distribution can then be chosen and used as the modeled 'estimate' of % ba/acre for that cell. Similarly, summary statistics such as the standard deviation or interquartile range of that distribution were calculated from the distri-

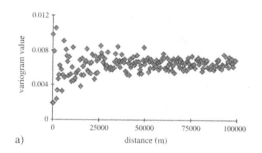
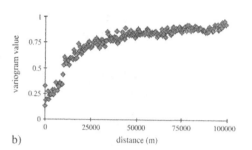

a)

b)

Figure 10.1. Spatial dependence as demonstrated by the sample variogram of the raw data (a) and sample variogram of the normal-scored data (b) for white pine.

bution of simulated values to describe the variation associated with the % ba/acre estimates for each cell. The estimate of each species' presence or absence was provided by indicator kriging. An indicator transform divides the data into two classes—either above or below a designated cutoff value; in this study 0% ba/acre. Indicator kriging calculates for each cell an estimate of the probability that it falls above or below the cutoff value. The output data set thus indicated the probability that sugar maple occurred at each location. Although this output does not include what can be strictly considered an uncertainty value, a probability of occurrence value can be effectively used to choose that cutoff that reflects the user's preferences for errors of omission vs. comission in the identification of areas of species occurrence.[1]

One advantage of indicator kriging is that it makes no assumptions about the distribution of the data. Sequential Gaussian conditional simulation, on the other hand, does assume that the data are normally distributed and stationary, and must be used more carefully (Deutsch and Journel, 1992). All species exhibited a proportional effect, with areas of high mean corresponding with areas of high local standard deviation, indicating a lack of global stationarity. The data were, however, locally stationary. As mentioned previously, sgCS assumes the data are multinormal. The data were transformed to be univariate normal, and were checked for bivariate normality by comparing the sample indicator variograms at each decile against what their bivariate normal equivalent would be (Deutsch and Journel, 1992). In each case, the species were found to be very close to bivariate normal. Higher moment normality was assumed.

Indicator kriging and sgCS were run for each species, using models derived from the appropriate sample variograms. The estimation parameters of cell size (3,000 m), search radius (10,000 m), and minimum:maximum number of points used (1:16) were taken directly from the results of the previous study. A stabilization check was run to determine the number of sgCS simulations necessary. Many simulations were run, and statistics were calculated from successively higher numbers of those simulations to determine at what

---

[1] All the methods used are described in Rossi et al., 1993, and Isaaks and Srivastava, 1989; the analysis was performed using GSLIB routines (Deutsch and Journel, 1992) with some additional routines written by R.E. Rossi. Srivastava (1994) provides a good description of stochastic modeling/simulation techniques.

Table 10.1. The Percent Variation Explained by the Spatial Dependence in the Sample Variograms

Species	Indicator Variogram	Normal-Scores Variogram
Beech	46	76
Black oak	37	75
Chestnut oak	44	82
Hemlock	57	66
Red maple	32	35
Red oak	38	64
Sugar maple	32	97
White oak	50	79
White pine	37	84
Yellow birch	39	81

point any additional simulations no longer substantially changed the summary statistic. For the species and cells examined, this occurred at roughly 80 simulations. To allow for some differences between species, 100 simulations were run for each species.

RESULTS AND DISCUSSION

With the exception of red maple, all species examined demonstrated substantial spatial dependence in the sample variogram of normal-scored data, with 64 to 97% of the variation explained by the visible structure and capable of being modeled (Table 10.1). In some species, much of that spatial dependence was contained in a very long-range trend of about 100,000 meters. White oak, black oak, chestnut oak, and beech all fell into this category. The rest of the species appeared to split the bulk of the explained spatial dependence over two ranges. For red oak, this was both a short- and medium-range pattern (12,000 and 40,000 m); sugar maple a very short- and a long-range pattern (2,100 and 60,000 m); and yellow birch, a medium- and long-range pattern (19,000 and 80,000) (Figure 10.2 and Table 10.2). The spatial dependence exhibited in the sample indicator variograms was much less, ranging from only 32 to 57% of the variation explained (Figure 10.3 and Table 10.3). This is not ideal for estimation and suggests the necessity for further refinement to decrease the uncertainty associated with those estimates. In these situations, other parameters, such as the size of the search radius, become a dominant effect in the interpolation and need to be even more carefully assessed. In general, however, where spatial dependence exists, incorporating that information into the model should improve the estimate.

The output from sgCS creates a frequency histogram of possible values for each cell. From this distribution, any summary statistic can be calculated for use as the 'estimate' and for the measure of uncertainty, depending on the objectives. Choosing the mean estimate, as expected, created a noticeably smoothed data set, similar to the output from ordinary kriging (with the exception that a robust uncertainty term such as the standard deviation could now be calculated with it). Choosing the median estimate data set in each case did maintain the sample histogram and did not noticeably smooth the data. And it created a data set that retained more

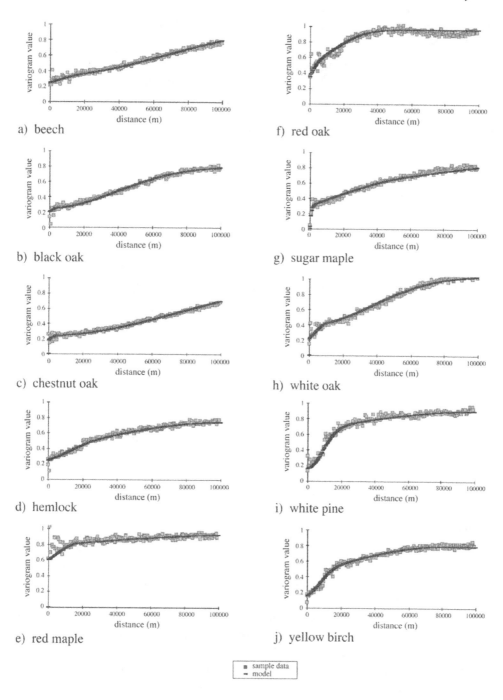

Figure 10.2a–j. Sample variograms from the normal-scored data and models for each of the 10 species. These models were used in the sgCS.

of the local variation. The median estimate, however, did provide a very low estimate of % ba/acre values, typically well below what was expected from previous knowledge. As a result, the 70th percentile was chosen as the 'estimate' of beech % ba/acre. It contained many of

Table 10.2. Isotropic Function and Parameters Used to Define the Model Used for Each Species in sgCS

Species	# Structures	Nugget Effect	Function	Range (m)	Component
Beech	2	.24	spherical	20000	.09
			Gaussian	160000	.67
Black oak	2	.20	spherical	5000	.05
			Gaussian	95000	.55
Chestnut oak	2	.18	spherical	5000	.06
			Gaussian	190000	.76
Hemlock	2	.25	Gaussian	30000	.11
			spherical	100000	.38
Red maple	2	.60	spherical	18000	.18
			spherical	120000	.15
Red oak	2	.35	exponential	12000	.25
			Gaussian	40000	.36
Sugar maple	3	.03	exponential	2100	.27
			spherical	60000	.19
			spherical	220000	.50
White oak	2	.22	spherical	11000	.18
			Gaussian	90000	.64
White pine	3	.16	Gaussian	21500	.48
			spherical	100000	.26
			spherical	140000	.10
Yellow birch	3	.15	exponential	6000	.03
			Gaussian	19000	.25
			spherical	80000	.36

the characteristics of the sample data in terms of the histogram and local statistics, as well as more closely maintaining the county averages. As the measure of uncertainty about this value, the middle two-thirds of this distribution was chosen.

The results create several possible output data sets. Plate 4 illustrates the results for beech, using four data sets to represent the species. Part (a) shows the estimated probability of beech occurrence, as calculated from indicator kriging. Part (b) is the 70th percentile value of the 100 sgCS simulations, representing the chosen estimate of beech in % ba/acre. The uncertainty associated with that estimate is described in (c) and (d). It was chosen to be a percentile range capturing approximately two-thirds of the distribution. Part (c) expresses the minus variation, or that distance in % ba/acre values between the 70th percentile and the bottom of that range (the 17th percentile), and part (d) expresses the plus variation (the difference between the 83rd and 70th percentiles). The minus variation map represents where and how much the real value is likely to be below the estimate presented, and the plus variation map depicts where and how much the real value is likely to be above the estimate.

The results of sgCS identify where concentrations of each species occurs much more specifically than county summaries. When species are relatively rare, as yellow birch and white pine are in Pennsylvania, or patchy, as red oak is, a summary at county level smooths the information enough to hide completely all the local detail as to where concentrations of

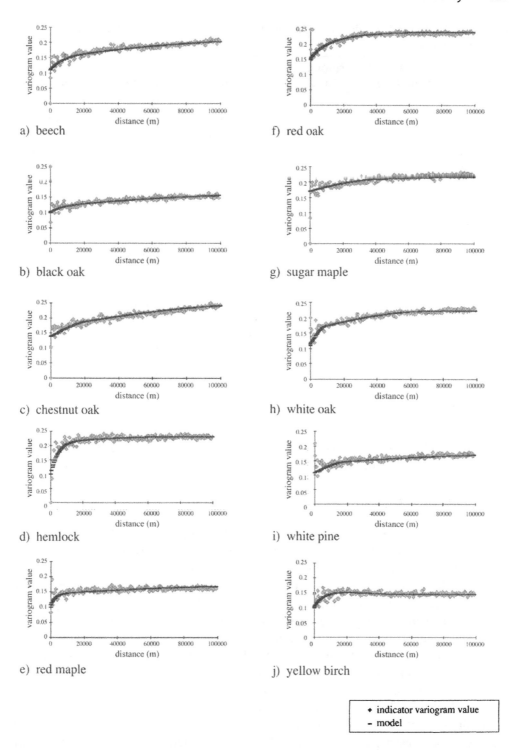

Figure 10.3a–j. Sample indicator variograms and models for each of the 10 species. These models were used in the indicator kriging (IK).

Table 10.3. Isotropic Function and Parameters Used to Define the Models Used in the Indicator Kriging

Species	# Structures	Nugget Effect	Function	Range (m)	Component
Beech	2	.11	exponential	20000	.04
			spherical	120000	.055
Black oak	2	.10	exponential	20000	.02
			spherical	120000	.0375
Chestnut oak	2	.11	spherical	21000	.025
			spherical	120000	.06
Hemlock	2	.08	exponential	13500	.09
			spherical	80000	.015
Red maple	2	.08	exponential	90000	.035
			spherical	120000	.02
Red oak	2	.15	exponential	30000	.07
			spherical	70000	.02
Sugar maple	2	.17	spherical	40000	.03
			spherical	100000	.015
White oak	2	.11	spherical	10000	.05
			spherical	70000	.06
White pine	2	.085	spherical	20000	.025
			spherical	120000	.025
Yellow birch	2	.08	exponential	20000	.05

these individual species occur within the counties. The results of sgCS highlighted rather than smoothed over this detail.

Estimated Data Sets and Associated Uncertainty

Plate 4 illustrates four output data sets that together provide useful information about species occurrence and distribution. The probability that a species occurs is a unique and useful data set. Species occurrence is often a first decision criterion in land management, particularly in situations where it is not yet known what density of a particular tree species is required before a wildlife species considers it suitable habitat, or before a pest finds conditions favorable. By reporting a probability of occurrence for each cell, the user is no longer faced with a fixed final map based on preferences assumed at the time of model creation, but can use any probability as the cutoff to create a species presence/absence map depending on the objectives of the task at hand. For example, if a particular insect is known to live in hemlock forests and the objective is to limit the search to only those areas where there is a high probability of finding suitable conditions, we might set the cutoff at the probability level of >= 0.8 to create a map of the area of interest. If, however, we are most interested in not missing any areas where the insect occurred, we might set the probability level for forest much lower, say >= 0.4.

The results of sgCS are particularly useful when there is interest in the distribution of densities at which a species occurs. For example, if it is known that a certain predominance of a chestnut oak is required for use by bear as permanent territory, but that a lesser dominance of oak is possible if certain other conditions also exist, an estimate of the % ba/acre of oak for each area is useful information. Plate 5 is a map produced from the sgCS estimates for chestnut oak, illustrating where it occurs as a major (>40% ba/acre), moderate (20 to 40% ba/acre), and minor (1 to 20% ba/acre) component of the forest. This estimate is associated with a corresponding map depicting the uncertainty found in the conditional simulation. In Plate 6 the uncertainty is divided into three classes reflecting what the user considered to be acceptable, moderate, and unacceptable levels of uncertainty. Highlighted in red in Plate 6 are those areas where the uncertainty is greater than the level considered acceptable—those areas where the local variability of chestnut oak value exceeded the limits of resolution of the current sample design and intensity. Such a map illustrates those areas to target should more time and money become available to collect additional inventory data for chestnut oak, and which areas to first discount should another decision criteria become available.

The same % ba/acre data set(s) also can be used to create a forest-cover-type "map." The advantage of creating and maintaining estimated data sets for each individual species is that forest cover types can be uniquely defined to capture more accurately the habitat required for a particular study. For example, the user might define a SM/B type as those areas where the combined % ba/acre of sugar maple and beech amounts to more than 50% of the total ba/acre in the stand, and where no more than 20% of the rest is yellow birch. For this purpose, the summary statistic (whether mean, median, or any of the percentiles of each cell's simulated distribution) would be chosen specifically for the intended purpose in the same way the different levels of probability were chosen when mapping species presence/absence from indicator kriging estimates. If the objective is to reduce the error of commission (i.e., classifying areas as sugar maple/beech (SM/B) in the estimated map that really are not SM/B), then using a percentile at the lower end of each cell's distribution would be more desirable. If, however, the objective is to reduce the error of omission (i.e., reduce the possibility of missing areas that do contain SM/B), then using a higher percentile from the distribution, such as 70%, would be more desirable (Plate 7).

This last application of the data is an ideal one, because it assumes that the joint distribution structure, i.e., those relationships of distribution between different species, are maintained in the output data set. In reality, however, in this study only the spatial structure is explicitly modeled and maintained. Work is ongoing to investigate ways to incorporate more explicitly these two forms of correlation or structure so commonly found in forested ecosystems (Moeur and Hershey, 1999).

Some kind of assessment of the output data sets is also important if the data sets are to be really useful. How well does the output data set portray the characteristics of the phenomenon it was designed to capture? If it was intended to estimate current land cover as observed on the ground, how well does it do that? If it was intended to capture local variability, how well does it do that? If it was intended to identify where there are higher concentrations of the resource, how well does it do that? This study considered three kinds of assessment:

1. assessment for the performance of the models and interpolation methods—to make sure they have not been biased or misused.
2. assessment for the sensitivity of the output to small changes in model parameters—to get some idea of how easily one might have come up with a different answer from the same data.

3. assessment of the relationship to ground values. And here, there are two parts:
   a. how well does it maintain the characteristics of the sample data we are interested in?
   b. how does it compare to actual values on the ground—do they fall within the level of uncertainty given?

Some of this assessment was performed on the final output, to help qualify the data set as to what it does or does not tell us about the phenomenon. For example, a check that compares the original joint distribution structure is appropriate to determine to what extent the individual species data sets can be queried together. Other assessments occurred throughout the analysis, such as the tests for model performance and sensitivity. For example, sgCS is designed to preserve closely the sample histogram and sample variogram. To check how well it performed, given the unknowns about the normality of the sample data, the histogram and sample variogram of several single realizations were compared with that of the sample data. In this study, the realizations output by sgCS recreated almost identically the sample variogram at lag distances up to 40,000 m, indicating that the sgCS, and the models and search criteria used, is performing up to expectations over those distances. At the larger lags, however, both the sample variogram and covariance remained close but not identical to that of the sample data, indicating that there is some structure in the data at the large lags that was not effectively incorporated in the simulation. Incorporating the structure that exists over these larger distances could be addressed using a two-stage simulation.[2]

Uncertainty and Variability in the Data

Local variability in the data is a reflection of the growth and distribution patterns of a tree species, local topography, and local land use histories. It affects the uncertainty of the estimate when % ba/acre values vary over shorter distances than the sampling intensity resolves. In this study, every tree species examined in Pennsylvania exhibited more local spatial variability than was completely resolved by FIA plots averaging 2.5 km apart. This unexplained variation was reflected in the sample variogram as a sometimes substantial nugget, particularly in the indicator variograms, and resulted in higher levels of uncertainty associated with the estimates in this study.

Spatial data sets, like nonspatial classifications, are themselves abstractions or generalizations of some spatial variation that is really there on the ground. Goodchild and Gopal (1989) describe two familiar examples: "The area labeled 'soil type A' on a map of soils is not in reality all type A, and its boundaries are not sharp breaks but transition zones. Similarly, the area labeled 'population density 1,000–2,000 sq. km.' does not in fact have between 1,000 and 2,000 in every square kilometer, or between 10 and 20 in every hectare, since the spatial distribution of a population is punctiform and can only be approximated by a smooth surface." But rarely can we, or do we want to, estimate down to a level of spatial detail such that each area is completely homogeneous, even among the few variables we may be interested in. The limits of the data, including sampling design and intensity, typically limit the level of resolution possible by interpolation. The inherent local variability of

---

[2] R.E. Rossi, personal communication, 1997.

the phenomenon is a major source of the uncertainty of the predicted value at each location when using only spatial information in the model and estimation.

One advantage to making the accuracy and uncertainty of a modeled estimate explicit is that it provides substantial information for effectively improving the estimate should the objective demand and should time and money become available to do so. As was observed with sugar maple, white pine, yellow birch, and hemlock, it may be possible to significantly improve the estimates of % ba/acre by refining the analysis. When subpopulations of a species have a significantly different pattern of spatial distribution, treating the populations separately in the interpolation will improve the final estimates. These populations may be described by regional land features or by some other defining characteristic (e.g., stand age for hemlock). For sugar maple, dividing the state into several broad ecological regions made a significant difference in the calculated sample variogram and thus in the final simulated estimates. Another important clue is previous knowledge about the species—e.g., that different ecological regions may be causing distinct spatial distribution patterns, or that different size class or age populations may have different spatial distribution patterns over the landscape. Hemlock is an example of the latter. As a result of past management practices that involved heavy harvesting of large hemlock for the tanning industry, today there are often relics of large individuals among a relatively wider distribution of smaller, younger trees that have grown up in the interim (Hough and Forbes, 1943; Powell and Considine, 1982).

The Appropriateness of the Variogram Model

The model describing the spatial relationship between neighboring locations is the critical element of any spatial estimation. The model is designed to match closely the spatial relationship observed in the sample data (e.g., the structure and dependence observed in the sample variogram), paying particular attention to those distances, usually the shorter ones, that are used in the estimation. However, when subpopulations of a species have a significantly different pattern of spatial distribution, calculating a single variogram for the entire area in effect captures more than one population within that single variogram and results in a model that is probably appropriate to neither population. In such situations, treating the populations separately in the modeling and estimation will certainly improve final estimates.

In this study, the % ba/acre estimates of four species were further refined. Sugar maple and the two rarer species, white pine and yellow birch, were divided into several different populations by region, and the process of variogram modeling and estimation was repeated for each region. With white pine and yellow birch, the regions were designed to target more specifically those areas in which the species was more dominant. With sugar maple, ecological subregions, at the section level, were used as the default regions for identifying potentially different populations of each species (McNab and Avers, 1994). In all three situations, the regionalized variograms were substantially different from the average variograms, suggesting that some improvement in the local applicability of the model should be possible by using this local spatial information. When sgCS was performed using the locally tuned models, it revealed subtle differences in the simulated results for yellow birch and white pine, and more substantial differences in the results for sugar maple, including a reduction in the uncertainty. Running a separate indicator model and indicator kriging on

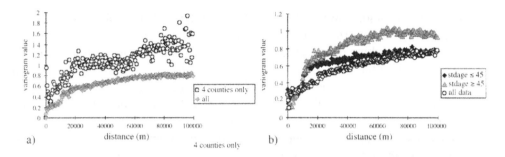

Figure 10.4. Sample variograms calculated for specified subpopulations of the sample data set. Separating subpopulations of (a) yellow birch by region and (b) hemlock by stand size resulted in distinctly different variograms both in range and sill.

the two regions also modified the final map; however, the effect was more subtle as a result of the smaller differences between the two models. A fourth species, hemlock, was refined based primarily on the suspicion from historical information that small and large stand-size classes may have different spatial distribution patterns. When variograms were calculated separately for these two populations (using a cutoff of 45 years), they were indeed substantially different in shape, sill, nugget, and range (Figure 10.4).

These data sets of individual species distribution do not contain any of the fine-scale forest/nonforest detail. If such information is desired, more detailed data sets describing the forest/nonforest land cover in Pennsylvania would be derived from a more intense point sample or the continuous but averaged data available from satellite imagery (e.g., Zhu, 1992). Such detailed data sets are then used as a 'mask' overlaid on any of the data sets of species distribution to provide a more realistic picture of where individual species currently occur amid the more spatially-detailed mosaic of forest/nonforest landcover (Plate 8).

CONCLUSIONS

Because we are working with spatially correlated phenomena that do exhibit spatial structure at this intensity of sampling and because we are looking for spatial output, geostatistical analysis and geostatistical interpolation techniques become very appropriate. In addition, geostatistical techniques offer ways to explore, organize, and summarize spatial patterns in the data that can provide clues to the variation and spatial behavior of the individual species under investigation. The estimated data sets output by indicator kriging and sgCS have the potential to be very useful. The resulting spatial data sets may be used simply as a focus for discussion; a catalyst for further analysis; or directly as a critical data set in other analyses, decisions, and models. Every species examined, with the exception of red maple, exhibited substantial spatial dependence in the sample variograms of the normal-scored data, suggesting that there is considerable benefit to incorporating that spatial structure in the estimation of species occurrence and importance from FIA data. There were also differences between species—each exhibiting some variety in spatial pattern and structure as well as in the amount of spatial dependence—and each may require different levels of additional fine-tuning, depending upon the objectives of the specific analysis and the time and expertise available. Many of the species examined for regional differences did exhibit such differences in the sample variograms, although the effect on the final simulated results did vary with the magnitude of that difference. Rare species, including those that occurred on less than 25% of the sample plots, were modeled more effectively by sgCS than by any method of local area averaging.

These geostatistical techniques make explicit the uncertainties associated with a modeled estimate in a form that can be incorporated when the data are used. This feature adds considerable utility and flexibility in the use of the resulting estimates, as the risk of errors of commission or omission can be specifically determined and manipulated to fit the current objectives. Recognizing the assumptions, limitations, and choice of errors in a map instead of blindly accepting the output allows the user to apply an estimated data set much more effectively and with a better understanding of its logical capabilities and precision. What the map is going to be used for strongly affects what is considered acceptable "error," and what types of error are most and least tolerable. As described above, one advantage of both the probability output of indicator kriging and the modeled distribution output of sgCS is that the information is available to create maps with an objective and desired error in mind.

Maintaining individual species information separately allows considerable flexibility in the use of species distribution data. Instead of being limited to previously defined fixed classes, forest cover types can be uniquely defined to capture more accurately the habitat required for a particular study. The potential also exists to use one or more of the species data sets as a decision layer in the interpretation of satellite imagery. The two data sets offer complementary information about the species composition that really exists on the ground.

Tree species exhibit more than just spatial autocorrelation; they may also exhibit some correlation with particular soils or topography or with reflectance data from satellite imagery. Based on the strength of that correlation, such data can be incorporated into several of the geostatistical techniques as ancillary or 'soft' information to improve the estimation of an individual species. Tree species also exhibit joint distribution relationships with other tree species. This fact is an important characteristic in any queries involving more than one species, yet these relationships are not necessarily maintained when each species is estimated separately as in this study. Other techniques, such as imputation (e.g., Moeur and Stage, 1995) are designed to maintain the joint distribution structure of variables, but do not take full account of the spatial structure. Investigation is currently underway to identify techniques that would better maintain both the spatial structure and the joint distribution/ attribute structure of the sample data, such as cosimulation, indicator simulation, or a combination of several techniques.

The techniques used in this study are not extremely time-consuming nor difficult to process, and could be easily extended to additional species, states, and variables. There is a high level of variance associated with these estimates of % ba/acre—in many locations this variance can be as much as the estimate itself. Nevertheless, the data sets provide a very descriptive picture of species distribution at the state level. In comparison to previous depictions of current species distribution from FIA data by summarizing at the county level, these methods provide a much more detailed picture of species occurrence and distribution. Estimates can probably be improved and variances diminished by additional investigation into and/or sampling of each species and by incorporating ancillary information for which we have more spatially detailed data; however, even the current estimates are informative and provide a useful basis from which to proceed.

REFERENCES

Alerich, C.L. *Forest Statistics for Pennsylvania—1978 and 1989,* Resource Bulletin NE-126. USDA Forest Service, Northeastern Forest Experiment Station, Radnor, PA, 1993, p. 244.

de Fouquet, C. and D. Mandallaz. Using geostatistics for forest inventory with air cover: An example. In *Geostatistics Tróia '92.* Soares, A. Ed. Kluwer Academic Publisher, Amsterdam, 1993, pp. 875–886.

Deutsch, C.V. and A.G. Journel. *GSLIB: Geostatistical Software Library and User's Guide,* Oxford University Press, New York, 1992, p. 340.

Goodchild, M. and S. Gopal. Preface, pp. xi–xv in *Accuracy of Spatial Databases,* M. Goodchild and S. Gopal, Eds., Taylor & Francis, New York, 1989, p. 290.

Goovaerts, P. *Geostatistics for Natural Resources Evaluation,* Oxford University Press, New York, 1997, p. 512.

Hansen M.H., T. Frieswyk, J.F. Glover, and J.F. Kelly. *The Eastwide Forest Inventory Data Base: Users Manual,* General Technical Report NC-151. USDA Forest Service, North Central Experiment Station, St. Paul, MN, 1992, p. 48.

Hough, A.F. and R.D. Forbes. The ecology and silvics of forests in the high plateaus of Pennsylvania, *Ecol. Monogr.* 13, pp. 299–320, 1943.

Isaaks, E.H. and R.M. Srivastava. *An Introduction to Applied Geostatistics,* Oxford University Press, New York, 1989, p. 561.

Liebhold, A.M., R.E. Rossi, and W.P. Kemp. Geostatistics and geographic information systems in applied insect ecology, *Annu. Rev. Entomol.* 30, pp. 303–327, 1993.

McNab, W.H. and P.E. Avers. *Ecological Subregions of the United States: Section Descriptions,* Administrative Publication WO-WSA-5. USDA Forest Service, Washington, DC, 1994, p. 267.

Moeur, M. and R. Riemann Hershey. Preserving spatial and attribute correlation in the interpolation of forest inventory data. In *Spatial Accuracy Assessment: Land Information Uncertainty in Natural Resources,* Lowell, K. and A. Jaton, Eds. Proceedings of the Third International Symposium on Spatial Accuracy Assessment in Natural Resources and Environmental Sciences, May 20–22, 1998, Quebec City, Canada, Ann Arbor Press, 1999, pp. 419–430.

Moeur, M. and A.R. Stage. Most Similar Neighbor: An improved sampling inference procedure for natural resource planning, *For. Sci.* 41(2), pp. 337–359, 1995.

Monmonier, M. *How to Lie with Maps.* The University of Chicago Press, Chicago, 1991, p. 176.

Mowrer, H.T. Spatially quantifying attribute uncertainties in input data for propagation through raster-based GIS, in *Proceedings of the 8th Annual Symposium on Geographic information systems,* Vancouver, BC, Canada, 1994, pp. 373–382.

Powell, D.S. and T.J. Considine. *An Analysis of Pennsylvania's Forest Resources,* Resource Bulletin NE-69. USDA Forest Service, Northeastern Forest Experiment Station, Broomall, PA, 1982, p. 97.

Riemann Hershey, R., M.A. Ramirez, and D.A. Drake. Using geostatistical techniques to map the distribution of tree species from ground inventory data, pp. 187–198 in *Modelling Longitudinal and Spatially Correlated Data: Methods, Applications, and Future Directions.* Lecture Notes in Statistics, Vol. 122, T.G. Gregoire et al., Eds., Springer-Verlag, New York. 1997, p. 402.

Rossi, R.E., D.J. Mulla, A.G. Journel, and E.H. Franz. Geostatistical tools for modeling and interpreting ecological spatial dependence, *Ecol. Monogr.* 62(2), pp. 277–314, 1992.

Rossi, R.E., P.W. Borth, and J.J. Tollefson. Stochastic simulation for characterizing ecological spatial patterns and appraising risk, *Ecol. Appl.* 3(4), pp. 719–735, 1993.

Samra, J.S., H.S. Gill, and V.K. Bhatia. Spatial stochastic modeling of growth and forest resource evaluation, *Forest Sci.* 35(3), pp. 663–676, 1989

Srivastava, R.M. An overview of stochastic methods for reservoir characterization, in *Stochastic Modeling and Geostatistics: Principles, Methods and Case Studies,* J.M. Yarus and R.L. Chambers, Eds., 1994, pp. 3–16.

Wackernagel, H. *Multivariate Geostatistics: An Introduction with Applications,* Springer-Verlag, Berlin, 1995, p. 256.

Zhu, Z. *Advanced Very High Resolution Radiometer Data to Update Forest Area Change for Midsouth States,* Res. Pap. SO-270. USDA Forest Service, Southern Forest Experiment Station, New Orleans, LA, 1992, p. 11.

Chapter 11

Realistic Spatial Models: The Accurate Mapping of Environmental Factors Based on Synecological Coordinates

Margaret R. Holdaway and Gary J. Brand

The spatial variability in the environment is responsible, in part, for the mosaic of vegetation types visible across a landscape. Our ability to appropriately manage forests or to accurately predict changes in them can be improved if we can accurately map environmental conditions. For example, a less abundant overstory species may be favored over more predominant species for sites with environments more conducive to the less abundant species. Maps may also show areas with highly variable environments and therefore alert managers to the possibility of a more variable response to activities in those areas. Unusual environmental conditions may also pinpoint locations that might harbor unique plant communities and therefore indicate areas worthy of additional field survey.

In this chapter we infer the spatial variability of environmental conditions via plant-based indices produced by the Method of Synecological Coordinates (MSC) (Bakuzis, 1959; Bakuzis and Kurmis, 1978; Brand, 1985; Gutiérrez-Espeleta, 1991; Gutiérrez-Espeleta, 1996). The basis for MSC is the time–honored notion that floristic composition of a site provides an indirect measure of the nature of the environment there.

MSC is an easily applied approach that computes site scores for four fundamental environmental factors—moisture (M), nutrients (N), heat (H), and light (L). Taken together, M and N represent the edaphic conditions, and H and L represent the climatic conditions at the site. Site scores are computed from indicator values for each species present. Each plant species has its own quadruplet of synecological coordinates representing its relative "requirements" for M, N, H, and L when competing with other plants. These integer values range from 1 (for dry, poor, cold, and dark conditions for M, N, H, and L, respectively) to 5 (for wet, rich, hot, and bright conditions for M, N, H, and L, respectively). An arithmetic mean of the coordinates for those species occurring together is called the site synecological coordinate for that factor.

Note that the mean is computed from ordered numbers; a computation that is usually recommended only if the interval or ratio scale applies (Stevens, 1946; Husch et al., 1972). An interval scale states that equal differences in the indicator value for M equate to a fixed amount of moisture. Because synecological coordinates are relative, not absolute, in plant indicator values (Gutiérrez-Espeleta, 1996), this is not the case. We justify the use of these problematic statistical methods from a pragmatic viewpoint. Stevens (1946) states: "In numerous instances it [treating ordinal scales as interval scales] leads to fruitful results."

This has been the case for MSC. It has proved useful in classifying plant communities (Kurmis et al., 1986) and in identifying floristic classes that differ in productivity (Brand and Almendinger, 1992). Therefore, we will apply MSC as it was originally developed by Bakuzis (1959).

Spatial variability of site synecological coordinates can be analyzed and mapped using geostatistical techniques. Geostatistics is the name given to a toolbox of statistical techniques that deal with spatially correlated environmental data (Journel and Huijbregts, 1978; Isaaks and Srivastava, 1989; Delhomme, 1978). The preceding references provide good background for those readers unfamiliar with the basic concepts of geostatistics. Geostatistical theory is based on the simple notion that measurements of continuous phenomena made closer together in space are often more alike or correlated than those made farther apart. The spatial variability or structure can be modeled via a mathematical function known as the variogram. Then kriging, an interpolation technique, uses information in the variogram to predict a grid of synecological coordinates at unsampled points.

Ordinary kriging, one of the simplest forms of kriging, assumes that the data points demonstrate local stationarity, i.e., they contain no significant trends over the reasonably homogeneous smaller regions designated by the interpolation search neighborhood (Isaaks and Srivastava, 1989). As a result, it is a fairly accepted belief that moderate trends in the data do not significantly affect ordinary kriging interpolations (Deutsch and Journel, 1992; Journel and Rossi, 1989). However, Laslett (1994) questions the "invocation of stationarity and the commonly repeated advice that trends rarely need to be modeled explicitly." Such skepticism suggests that a closer look at how to best model moderate trends in the data may be needed.

The different kriging methods vary in degrees of complexity and in their underlying assumptions. Several methods may yield a plausible version of the unknown reality. Users of geostatistics are faced with the task of choosing the most appropriate kriging method (or model) for their data. It is necessary to fit the appropriate models and then to compare their predictive capabilities. Criteria for comparing plausible models will not only include how close they are in making predictions, minimizing the bias and variability, but will also include whether they produce realistic results. Along with the already common statistical analysis of the residuals, there must be an assessment of how well the models describe the spatial features of the data (Haining, 1990). Such an assessment can be facilitated by a graphical comparison and evaluation of model predictions with properties of the data, making model sensitivity analysis a vital part of model validation.

We are investigating the application of geostatistical techniques to represent the local edaphic environment (via M and N synecological coordinates) of a contiguous block of forested land in St. Louis County, Minnesota. Our purposes here are: (1) to describe the spatial variation of M and N site synecological coordinates by three plausible geostatistical models and (2) to show the importance of extending model selection criteria beyond statistical criterion to include model sensitivity analysis.

METHODS

Study Area

The study site, 30 km north of Duluth (47°4′ N 92°3′ W; within NE 1/4 Sec. 24, T 53 N, R 14 W), Minnesota, is gently rolling with no obvious directional trends in the topography.

The predominant glacial advance was from northeast to southwest. Overall drainage in the region flows from the northeast to the southwest through a series of lakes and marshes formed by the glacial action. Soil parent material is red stony till and outwash deposited by the St. Louis sublobe or mixed sediments from the Rainy, Superior, and Wadena lobes (Cummins and Grigal, 1981).

Site Synecological Coordinate Data

To evaluate the spatial variation in M and N at this site, we established an 11×11 grid of points in a N-S, E-W direction across the study area. Grid points were 45 m apart. Ten of these primary grid points were randomly selected to provide auxiliary spatial information at close distance in the four cardinal directions. For these clustered grid points, two points were located 15 m to the south and west of the primary grid point and two points were located 5 m to the north and east of the primary point. An additional 24 grid points were located 31.8 m southwest of randomly selected primary grid points. Plant species present were tallied on 5×5 m plots centered on all 185 grid points. Locations of sample plots where plants were present are given in Figure 11.1. Plants present were assigned synecological coordinates developed for plants of Minnesota forests (Bakuzis and Kurmis, 1978), and site synecological coordinates were computed.

Modeling the Spatial Variability

The interpolation approach, kriging, consists of three parts: (1) calculating the experimental variogram from the data values, (2) modeling the variogram, and (3) using the modeled variogram equation to spatially interpolate data to unsampled points.

First, the experimental variogram is calculated from the data. Modeling the spatial variation began with fitting a theoretical variogram function to the experimental variogram. In addition to using different functions, kriging models can assume that the data exhibit isotropy, anisotropy, or trends. The isotropic model, the simplest of the three, assumes that the variogram function used to describe the spatial continuity varies by distance but not by direction. The anisotropic model assumes there are directional influences in the data; whereas the trend model is based on modeling trends in the data, which are not necessarily directional in orientation. Each model uses a different approach to account for important features of the data. In data showing moderate trends, it is judicious to evaluate all three models.

Once the spatial variation had been modeled, ordinary kriging was used to interpolate unsampled data points based on weighted local averaging of neighboring points. A contour map of the factor was drawn from data points interpolated over a fine grid using block kriging.

Before fitting the models, exploratory data analysis was used to examine the data for extreme values or outliers that may obscure the structure of the variogram or may severely alter interpolated values in their vicinity (Isaaks and Srivastava, 1989). Outliers can be classified as global (those observations that are substantially greater or less than all other observations regardless of their spatial location) or as local outliers (isolated values that are abnormal with respect to the local spatial configuration of values (Wartenberg, 1989; Haining, 1990). An observation will be classified as a potential global outlier if it is identified as an

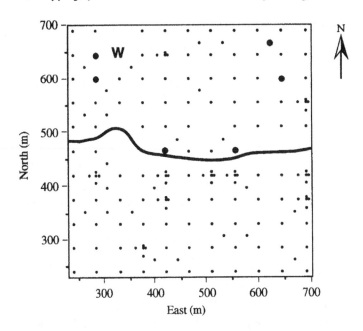

Figure 11.1. Sample plot locations (those observations occurring on a road or in a wetland are absent.) Global/local outliers removed for Moisture (•), and area of standing water (W), and the location of the east-west road are marked.

outlier on the box plot. Only those outliers that show sound physical reasons for their aberrant behavior from field records will be removed from the database (Isaaks and Srivastava, 1989). Local outliers will be detected by interpolating each observation from the surrounding observations and comparing the estimated value with the known value. All local observations that are obviously unlike their neighboring values (i.e., identified as outliers on a box plot of the residuals) will be removed.

Once global and local outliers were removed, the spatial structure was modeled using GSLIB (Deutsch and Journel, 1992). First, isotropic variograms were fit to the experimentally measured data using nonlinear least squares regression, weighted for the number of observation pairs in each distance class. Isotropic variograms for both factors (M and N) were calculated, using distance classes of 45 m (after a first class of 22.5 m).

Contours of the data showed evidence of directionality or possible larger-scale trends in the spatial variation of both environmental factors. Furthermore, the assumption of local stationarity may not be true. It was, therefore, appropriate to examine the data for anisotropy and trends.

Pairs of perpendicular directional variograms were calculated for angles from 0 to 90° (0° is north) in angular increments of 10° (with a ± 22.5° tolerance). The maximum and minimum range and the direction, ψ, of the maximum range were ascertained. The anisotropy ratio is the range in the major direction divided by the range in the minor direction and is a measure of the strength of the anisotropy. The M and N anisotropic models were subjectively fit by the linear function.

Ordinary kriging, used to interpolate the previous models, assumes that the data points are stationary, containing no significant trends over the spatial range. However, many environmental variables show local trends or even broader regional trends. To remedy the nonstationarity problem when the trend becomes too large, "universal" kriging was developed. "Kriging with a trend model" is the approach used in GSLIB to resemble universal kriging (Deutsch and Journel, 1992). Recall that even in the presence of moderate spatial trends, many researchers consider the simpler ordinary kriging model without trend to be sufficient. However, even if a simpler model appears adequate, there may be additional justification found in the model evaluation for using the more complex linear trend model.

To compare the results of these approaches, ordinary kriging (with isotropy and anisotropy) and kriging with a trend model were used on each factor to contour the data using 2×2 blocks at a resolution of 5 m. We produced contour maps of kriged values for all three models for the optimal search radius and number of observations (Spyglass, 1993).

Model Evaluation

Assessing the appropriateness of a model involves two components: cross validation statistics and model sensitivity analysis (spatial analysis of cross validation residuals and the model's interpretive ability). Not only are the cross validation statistics important, but the spatial features of the residuals are also important in evaluating model performance and identifying problem areas (Isaaks and Srivastava, 1989; Haining, 1990).

Cross Validation

The cross validation technique was used to statistically validate the three models. In this technique, each known data observation is temporarily deleted and the remaining neighboring points used to predict the deleted observation by kriging (Cressie, 1991). The prediction errors are calculated from the predicted minus actual values (Cressie, 1991) and characterized using common statistics. The aim is to simultaneously minimize the mean error (ME), median absolute error (MAE), and root mean square error (RMSE). Since the RMSE may be unduly influenced by a few values that are difficult to predict (Haining, 1990), the procedure can be weighted to favor commonly occurring conditions by including the mean absolute error (MAE) in the evaluation. Thus, three cross validation test statistics, which emphasize different attributes of the residuals, are given for each model. The ME should be close to zero and the RMSE and MAE should be minimized. These statistics were also useful in optimization of the search criteria (Isaaks and Srivastava, 1989), such as the search radius (SR) and number of observations within the search radius.

Model Sensitivity Analysis

Our approach to model sensitivity is to compare contour maps of the predicted values against a symbol map of the actual values. The symbol map provides a straightforward method to view the location, size, and shape of assorted spatial features. The approach can be narrowed in scope by comparing how each model interpolates only the dominant features (as indicated by data values above or below certain threshold values) with how those features are mapped by each model.

The spatial arrangement of fairly distinct areas of the actual observations when compared to the interpolated values can be used to indicate failure of a model to adequately

Table 11.1. Isotropic Variogram Function and Parameters Fit to Site Synecological Coordinates: Moisture (M) and Nutrients (N)

Factor	Function	Nugget Effect	Sill	Range (m)	r^2
M	Linear	0.011	0.023	343	0.90
N	Gaussian	0.019	0.118	599	0.99

account for important features of the mapped variable. Unreliable values will be found under certain conditions because of the inherent assumptions (such as directional influences) built into the model used. To help distinguish between likely model alternatives, a model sensitivity analysis of the data should suggest which model assumptions are the most essential to accurately reflect the prominent features of the data (Haining, 1990).

Kriging smooths out variation and reduces the extent and even strength of the very high and very low areas. However, it is important to identify which kriging models produce contour maps most closely honoring the observed high and low areas. Furthermore, if these extreme areas can be associated with certain environmental, edaphic, topographic, or glacial features, valuable information in evaluating the various models can be obtained.

RESULTS

Model Fitting

Site synecological coordinates (M and N) were computed from plots centered on 175 grid points (Figure 11.1). Plots at an additional 10 grid plots did not contain terrestrial vegetation. Five observations for M were identified as global outliers. Four occurred near the road or wet areas and were dropped. The fifth showed no irregularities on field records and was retained. An additional two M observations were classified as local outliers by their residuals and were dropped because of their effect on surrounding interpolations. No global or local outliers were found for N. To summarize, for M 169 plots remained; all 175 plots were used in the N analysis.

Site coordinates for the remaining plots depicts a fairly homogeneous study area that is moderately dry and slightly low in nutrients. Extremely dry or nutrient poor sites were not encountered within the study area. Histograms of M and N site coordinates were not skewed.

Model and parameters for the experimental isotropic variograms for M and N are given in Table 11.1. The experimental variogram for M was fit by the linear function with nugget effect. The slope was expressed in terms of the range and sill at about 350 m. A linear model, demonstrating a steep increase in semivariance with increasing distance, is indicative of a strongly nonstationary process (Haining, 1990). The model used here is characterized by a modest increase, which may reflect moderate nonstationarity.

The experimental isotropic variogram for N is best described by the Gaussian function with nugget effect. The Gaussian function has behavior indicative of a highly continuous phenomenon (Isaaks and Srivastava, 1989). However, its characteristic parabolic behavior near the origin may be an indicator of drift in the data or a regional trend (David, 1977). A study of anisotropy in the data or the trend model should confirm this.

Table 11.2. Anisotropic Variogram Function and Parameters Fit to Site Synecological Coordinates: Moisture (M) and Nutrients (N)

Factor	Function	Nugget Effect	Sill	Max R (m)	Min R (m)	k [a]	ψ [b]	r^2
M	Linear	0.011	0.028	850	362	2.35	120°	0.91
N	Linear	0.014	0.082	850	296	2.87	70°	0.91

[a] k: anistropic ratio = (max range/min range).
[b] ψ: direction of maximum variation (degrees E of N).

Differences in variation due to directional orientation were identified for both factors. The direction of maximum continuity for M occurs at 120°; for N, maximum continuity was evident in directional variograms oriented at 70°. The slopes of the major and minor axes were fit by the linear function; parameters for the anisotropic model are presented in Table 11.2. The slope is expressed in terms of the sill and range.

Model Evaluation

Cross Validation

Results of the cross validation are summarized in Table 11.3. ME reveals that the unbiased condition is clearly met; MAE indicates that on the average, for M, predictions will be within 0.100 units of the true value. Predictions for N show more variability. The model that simultaneously minimizes ME, MAE, and RMSE is the linear trend model for M and the anisotropic model for N.

Because kriged estimates produce a smoothed surface, maps will show conservative estimates of highs and lows. However, some models, with superior assumptions for the prevailing environmental conditions, will produce less smoothing than other models. The magnitude of the smoothing effect of each model is reflected in the range of kriged estimates. The observed values had a range of 2.05 to 2.75 for M and 1.94 to 3.00 for N. For both M and N, the linear trend model shows the least smoothing of the minimum and maximum values (Table 11.3). However, for N the linear trend model is not the best statistical model.

Model Sensitivity Analysis

To identify those areas that are not well summarized by any given model, we compared prediction maps of the three models being evaluated (Figures 11.2a–c and 11.3a–c) with a modified symbol map of the actual observations (Figures 11.2d and 11.3d).

For M, two areas of higher values along the southern border and one along the eastern border (Figure 11.2d) show an apparent directional influence to the southeast. The anisotropic model (Figure 11.2b) appears to highlight these two high areas and imposes their directional tendency on all other features of the map, causing an elongation of contours in the SE-NW direction that is often inappropriate. It is therefore judged the worst model. The isotropic model (Figure 11.2a) and linear trend model (Figure 11.2c) produce somewhat similar results, but on closer examination, the linear trend model performs better in render-

Table 11.3. Cross Validation Statistics for Site Synecological Coordinates: Moisture (M) and Nutrients (N). (ME is mean error, MAE is mean absolute error, SR is the search radius, and NObs is the number of observations within the search radius.)

Model	Function	SR	NObs	ME	MAE	RMSE	Min/Max est[a]
			M				
Isotropic	Linear	130	12	0.000	0.100	0.124	2.24–2.56
Anisotropic	Linear	260	16	0.000	0.100	0.122	2.25–2.55
Linear trend	Spherical	130	16	0.000	0.098	0.122	2.19–2.65
			N				
Isotropic	Gaussian	130	18	−0.001	0.126	0.154	2.15–2.77
Anisotropic	Linear	260	14	−0.000	0.118	0.144	2.13–2.81
Linear trend	Spherical	260	14	0.001	0.123	0.148	2.09–2.81

[a] Min/max est = minimum and maximum of estimated site synecological coordinates.

ing the distinct localized trends. Two small areas, each marked by two low observations (M ≤ 2.30) on the symbol map, are more accurately accounted for by the linear trend model. These local minima are smoothed over by the isotropic model. Although differences among the three models in the statistical analysis appear marginal for M, the model sensitivity analysis, which attempts to accurately reflect the observed spatial features of the data, clearly favors the linear trend model.

The most striking feature of the N symbol maps (Figure 11.3d) is an elongated area in the NE quadrant of very low N (≤ 2.20). The isotropic model is incapable of predicting the feature, and the linear trend model barely differentiates it. The anisotropic model properly delineates the feature as well as two other elongated anomalies of moderate N (2.40 to 2.60) in the northern portion of the study area. Location, size, and shape of the large irregular area of high N (≥ 2.70) in the lower portion are very poorly predicted by the isotropic model. The anisotropic model also estimates this area the best. The body of the high N area for the anisotropic model appears to run NE-SW, similar to that found in the elongated low nutrients feature. The directional variogram determined the overall average major axis to lie slightly more southerly (i.e., 70°) than the NE-SW direction. Again, the spatial analysis of model sensitivity has matched the cross validation results, but has magnified the differences.

Final shaded contour maps of the selected models are given in Figure 11.4. Since the directional influences for the localized wet areas of M differ somewhat in orientation from each other, the linear trend model performs best in depicting these features. For N, with roughly an NE-SW trend evident over much of the area, the directional model is the most appropriate model. The anisotropic model is necessary to give greater emphasis to observations within that orientation; otherwise, certain very distinct features of the data are ignored.

DISCUSSION

The approach presented has jointly emphasized (1) analyzing the statistical properties of the interpolations and (2) evaluating the sensitivity of the models to interpret the data properties before making the final modeling decision (Haining, 1990). The first was expressed in the clear statistical terms of the ME, MAE, and RMSE. The second, although more

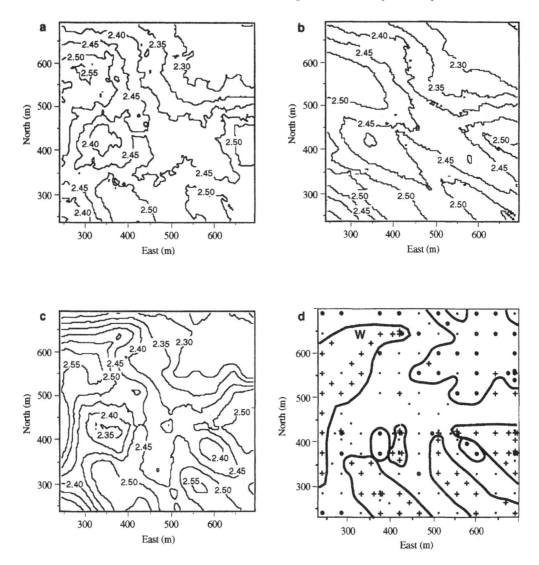

Figure 11.2. Contour maps of interpolated Moisture values for the (a) isotropic model, (b) anisotropic model, and (c) linear trend model compared with a (d) symbol map of actual data (+ is M ≥ 2.50, • is M ≤ 2.30, and · is all other M).

subjective, demonstrated the necessity of visually evaluating model predictive performance. By choosing to examine only the prominent spatial features of the data, the second emphasis has clearly defined problem areas with certain models that were not uncovered by the statistical analysis. Each model applies different simplifying assumptions (such as directional influences or trends) to describe the spatial characteristics of the data. By imposing these assumptions on the entire data set, they may either assist or hinder the model from accurately portraying the features of the data.

When a model accurately differentiates the prominent features of the sampling space, it may reveal possible explanations for those features—relating them to specific landscape

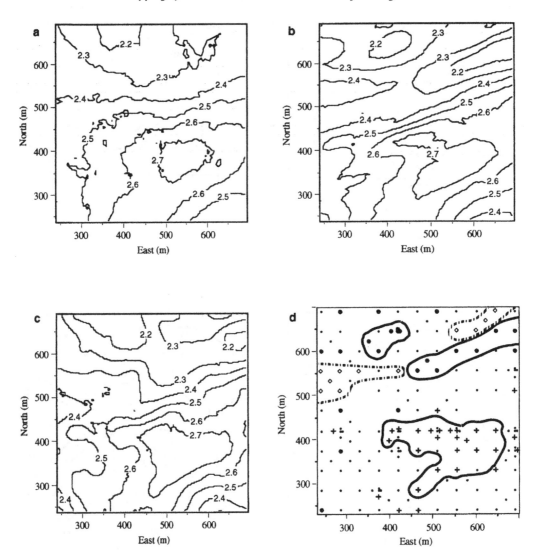

Figure 11.3. Contour maps of interpolated Nutrients values for the (a) isotropic model, (b) anisotropic model, and (c) linear trend model compared with a (d) symbol map of actual data (+ is N ≥ 2.70, ● is N ≤ 2.20, and · is all other N). Solid lines delineate areas of predominantly high or low N values. Dotted lines identify two isolated areas with values from 2.40 to 2.60 (◊).

features, geological conditions, or underlying physical processes that operate in defined directions. For N, the presence of elongated features of similar values oriented nearly in an NE-SW direction (Figure 11.4b) corresponds to the direction of glacial action in the region.

For M, compare the linear trend model interpretation (Figure 11.4a) with the topographic relief of the study area (Figure 11.5). The four areas of higher moisture all seem to emanate from higher regions of the plateau area (elevation above 1480 ft) in the southwest quadrant. One logical explanation would be the action of heavy rains and erosion on the forest landscape. Runoff from the higher elevations would begin to concentrate into broad, shallow channels separated by slightly higher areas that have been less affected by the runoff and

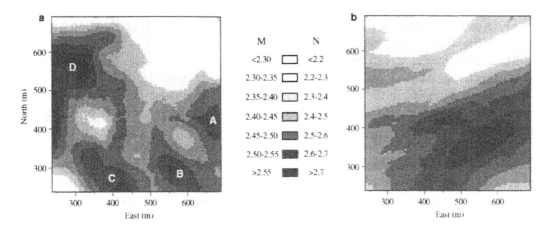

Figure 11.4. Shaded contour maps (a) Moisture using the linear trend model and (b) Nutrients using the anisotropic model. Four high Moisture areas are delineated (A, B, C, and D).

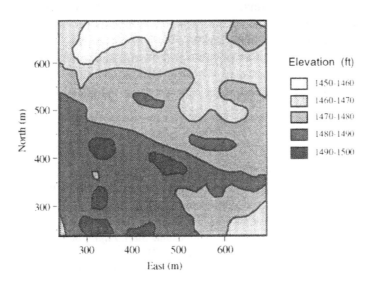

Figure 11.5. Topography of the study area.

erosion. Steepness of the slope and soil permeability will also determine how much of the runoff infiltrates the ground and how much flows off over the surface toward more low-lying areas. In less undulating areas, the influence of these drainage systems may not be as discernible in the landscape, but their presence should still be noted by the presence of more moisture-dependent plants.

There is definitely evidence of such influences. For reference, four high moisture areas have been delineated by A through D in Figure 11.4a. Runoff from area A empties into a depression to the southeast of A; runoff from areas B and C drains into a second depression to the south of B. Most of the higher elevations on the plateau could conceivably drain into area C. The drier island at (350,420) of Figure 11.2c corresponds to a high elevation area. The topography of the area indicates a fairly rapidly changing elevational gradient from this high mound northward to a low area (in the northwest quadrant) of standing water, a slope corresponding to the high moisture area D.

Three of the four features, representing areas of high moisture (Figures 11.2d and 11.4a), show directional channels developing toward the southeast. However, taken together, they seem to radiate out from the high elevation area in the southwest quadrant. Hence, the linear trend model performs better than the directional model in depicting all of the higher moisture features.

We might expect the northeastern quadrant, with decreasing elevation, to be wetter than it is. However, in addition to topographic position, soil texture strongly influences soil moisture. Soil texture was tabulated for a grid of points, approximately 135 m apart. In the northern areas of low N and often low M, sandy soils tended to occur. In the large area of high N (and moderate to high M) in the southeast quadrant, finer textured soils were observed.

The moderately large nugget effect for both M and N indicates the influence of localized sources of variation. However, the large ranges and the large-scale patterns evident in Figure 11.4 suggest that broader trends are also operating for both factors, such as the orientation of glacial features, water drainage patterns, and soil texture. Synecological coordinates have reduced the complexity of the spatial variation so that microsite variation in soil properties no longer dominates, but rather broader trends are revealed.

The accurate mapping of site synecological coordinates has helped capture the pattern of environmental variation. Derivation of the most appropriate map requires the use of the more subjective sensitivity analysis along with the usual statistical evaluation.

REFERENCES

Bakuzis, E.V. Synecological Coordinates in Forest Classification and in Reproduction Studies, thesis presented to the University of Minnesota, St. Paul, in partial fulfillment of the requirements for the degree of Doctor of Philosophy, 1959.

Bakuzis, E.V. and V. Kurmis. Provisional List of Synecological Coordinates and Selected Ecographs of Forest and Other Plant Species in Minnesota. *Staff Paper Series No. 5,* University of Minnesota, Department of Forestry, 1978, pp. 1–31.

Brand, G.J. Environmental Indices for Common Michigan Trees and Shrubs, Research Paper NC-261, USDA Forest Service, 1985, pp. 1–5.

Brand, G.J. and J.C. Almendinger. Synecological Coordinates as Indicators of Variation in Red Pine Productivity Among TWINSPAN Classes: A Case Study. Research Paper NC-310, USDA Forest Service, 1992, pp. 1–10.

Cressie, N. *Statistics for Spatial Data,* John Wiley & Sons, Inc., New York, 1991.

Cummins, J.C. and D.F. Grigal. Soils and Land Surfaces of Minnesota, 1980. *Soils Series No. 110,* University of Minnesota, Agri. Exp. Stn., 1981, pp. 1–59.

David, M. *Geostatistical Ore Reserve Estimation,* Elsevier, Amsterdam, 1977.

Delhomme, J.P. Kriging in the hydro sciences. *Adv. Water Resour.* 1, pp. 251–266, 1978.

Deutsch, C.V. and A.G. Journel. *GSLIB Geostatistical Software Library and User's Guide,* Oxford University Press, New York, 1992.

Gutiérrez-Espeleta, E.E. Tropical Forest Site Quality Assessment: An Approximation in Costa Rica, thesis presented to the Iowa State University, Ames, in partial fulfillment of the requirements for the degree of Doctor of Philosophy, 1991.

Gutiérrez-Espeleta, E.E. Some thoughts on ordination techniques, ecological study units and the method of synecological coordinates. *J. Theor. Biol.* 179, pp. 13–23, 1996.

Haining, R. *Spatial Data Analysis in the Social and Environmental Sciences,* Cambridge University Press, Cambridge, 1990.

Husch, B., C.I. Miller, and T.W. Beers. *Forest Mensuration,* 2nd ed. The Ronald Press Company, New York, 1972.

Isaaks, E.H. and R.M. Srivastava. *An Introduction to Applied Geostatistics,* Oxford University Press, New York, 1989.

Journel, A.G. and C.J. Huijbregts. *Mining Geostatistics,* Academic Press, New York, 1978.

Journel, A.G. and M. Rossi. When do we need a trend model in kriging? *Math. Geol.* 21, pp. 715–739, 1989.

Kurmis, V., S.L. Webb, and L.C. Merriam, Jr. Plant communities of Voyageurs National Park, Minnesota, U.S.A., *Can. J. Bot.* 64, pp. 531–540, 1986.

Laslett, G.M. Kriging and splines: An empirical comparison of their predictive performance in some applications. *J. Am. Stat. Assoc.* 89, pp. 391–400, 1994.

Spyglass, Inc. Transform, Ver. 3.01. Spyglass, Inc., Champaign, IL, 1993.

Stevens, S.S. On the theory of scales of measurement, *Science* 103, pp. 677–680, 1946.

Wartenberg, D. Exploratory spatial analyses: outliers, leverage points, and influence functions, in *Proc. Spatial Statistics: Past, Present, and Future.* Institute of Mathematical Geography, Syracuse University, 1989, pp. 133–162.

Chapter 12

Discrete Polygons or a Continuous Surface: Which is the Appropriate Way to Model Forests Cartographically?

Kim Lowell

BACKGROUND AND CONTEXT

Spatial Interpolation

Spatial interpolation is a technique whose basic concept has existed since long before digital geographic information systems (GIS) were created. It is used to obtain estimates over an entire surface for a given variable that has only been sampled at isolated points. A common example is estimating rainfall and/or temperature over an entire region for which weather data were collected at isolated meteorological stations. But even this seemingly simple and straightforward example masks a number of potential problems with using interpolation in such a situation. All are related to two underlying implicit assumptions that must be satisfied if the resulting interpolation is to be a valid representation of the phenomenon of interest.

First, the phenomenon being interpolated must truly be continuous. To illustrate this point, suppose that one wishes to model rainfall for a region only during a summer period of hot, humid weather. In such conditions, it is likely that a large amount of the precipitation that falls will be convective in nature—a type of precipitation that produces considerable amounts of rain in spot locations rather than a smooth gradient of rainfall over a large region. In this case, rainfall for the summer period may not be an interpolable phenomenon due to the seemingly random spatial nature of convective precipitation.

Second, while the initial necessary condition for interpolability is a continuous phenomenon, the second important factor is that the phenomenon being studied must be sampled at a density appropriate for its continuous nature to manifest itself. Suppose that in another region, all precipitation is orthographic—a type of precipitation that is strongly related to the topography of a region. Further suppose that in the region being studied, storm fronts arrive from the west and, as they move to the east, encounter a series of mountain summits of increasing elevation evenly spaced 10 km apart. It is reasonable to believe that the western slopes of these mountains will receive increased rainfall as one moves toward a summit. Moreover, the amount of rainfall will fall off sharply as the summit is crested, given that a storm rises in contact with a mountain slope, cools, releases all possible precipitation, and moves toward the next higher mountain where the process repeats itself. In this hypothetical case, if weather stations are spaced, say, 40 km apart, the resulting data for rainfall may not lend themselves to being interpolated even though the phenomenon being studied is known to be continuous. A better approach might be a regression model in which the amount of rainfall is estimated as a function of elevation and west- or east-facing slope.

Thus the ability to conduct reliable spatial interpolation depends both on the nature of the phenomenon being studied, and also the way in which it has been sampled.

Forest Parameters

Traditionally, forest maps are composed of nonoverlapping polygons having nonoverlapping appellations within the classification system used. Yet recently—particularly with the advent of GIS and associated technologies—interpolation has begun to be used to produce continuous surfaces from forest sample points for various forest parameters.

And, indeed, there may be reason to believe that at some scale, forest parameters are interpolable phenomena. Given a single forest stand, for example, it may be that "a lot" of its volume is concentrated in a relatively small area with gradations toward adjoining forest stands. Some researchers (Joy and Klinkenberg, 1994) have noted the tendency for forest polygon cores to be the most precise—i.e., repeatable by several interpreters—part of a polygon, suggesting that forest parameters may show a gradient from polygon center to polygon edges. Furthermore, it has been shown that there is a weak tendency for forest volume at a location to be related to the distance the location is from a cartographic boundary and the forest type on the other side of the boundary (Lowell, in press). Such information has led to at least one representation of spatial certainty for various forest types based on the idea that the polygon centre is the part of the polygon which is most likely to truly be the forest type labeled—an interpolation of certainty (Lowell, 1994a). And between forest stands of the same type or even different forest types it is possible that forest parameters are interpolable. A pure spruce stand in northern Quebec may contain less volume than the same type of stand in southern Quebec. Similarly, a high volume stand in one location may suggest that surrounding stands will also have high volumes due to favorable conditions related to underlying factors such as soil type and water availability.

The converse may also be true, however. Consider that the spatial structure of within-stand variance is not known and would seem to be strongly related to the forest sampling unit employed—a smaller sample unit giving a larger variance than a larger unit. Furthermore, in certain stands the spatial variability of tree locations and the characteristics of the sample units employed may render the spatial structure of the variance seemingly random—i.e., noninterpolable. Similarly, if a forest has been subjected to periodic disturbances such as fires or severe wind, the characteristics of one stand may provide absolutely no indication of the characteristics of nearby stands. And even different forest stands within the same forest type may show no gradation of productivity. Over a long range—i.e., northern to southern Québec—it may not be possible to have the same forest type appear over such a range due to biological carrying capacities of various sites. And even over a smaller region, the productivity of the stands of a given forest type may be more related to local ecophysiographic conditions than to merely a gradation of productivity.

Objectives

The purpose of this chapter is to examine the question of whether or not forest volume can be treated as a continuous—i.e., interpolable—phenomenon in the boreal forest of Quebec. This question will be addressed from a number of perspectives as it concerns volume among forest stands of different forest types, different forest stands in the same forest type, and within individual forest stands. In addition to assessing whether or not volume is interpolable at any of these scales, the implications of conducting interpolation when it is not warranted for these conditions will also be evaluated and presented.

STUDY AREA, DATA, AND METHODS

Two real-world forestry data sets were employed in this study. Both contain the per hectare wood volumes estimated for a collection of 0.04 ha sample plots located in Montmorency Forest—the research forest of Laval University located 80 km north of Quebec City. The dominant vegetation on Montmorency Forest is the boreal forest dominated by balsam fir (*Abies balsamea* [L.] Mill.) and to a lesser extent spruce (*Picea* spp.) and white birch (*Betula papyrifera* Marsh.) (Bélanger et al., 1988).

The first data set is composed of 92 sample plots distributed in a semi-random stratified sample throughout the 1,500 ha study area. These plots had been established as part of normal management activities to assess the harvestable volume on the forest. They were therefore located in mature stands or those approaching a harvestable volume. The density of these plots (1 per 16 ha or 1 plot every 400 m on average) is approximately 20 times as dense as forest inventories normally taken in Quebec (Beaulieu and Lowell, 1994).

The second data set contains 81 plots located within a 16 ha square in Montmorency Forest. These plots were placed on a 9-by-9 grid with plots being located 50 m apart. The 16 ha grid was located intentionally such that a number of forest types covering a range of forest conditions were included in the sample. This included clear-cuts, partial and total windthrow, and mature and developing forests.

Two types of analyses were conducted on each data set in its entirety and a series of subsets. These subsets will be described subsequently.

First, to assess the interpolability (or "continuous-ness") of volume, spatial autocorrelation was measured among the plots of a given data (sub)set (see Table 12.1). This was assessed by using both Moran's *I* and Geary's *c* (Griffith, 1985) and the assumption that sample plots were adjacent if, in a Voronoi diagram constructed around the plots, their Thiessen polygons touched. Note that this method—i.e., immediate adjacency—for assessing spatial autocorrelation was chosen intentionally over distance-based methods such as semivariograms (see, for example, Burrough 1986). This study was not intended to produce "optimal" volume surfaces. Rather, the central question in this study is whether or not the forest can be modeled as a continuous surface. If this is the case, then the volume measured at one point should provide an indication of the volume at a neighboring point—i.e., it is expected that high values will have a tendency to cluster in the same area, as will low values. This is the same as saying that positive spatial autocorrelation will be present among adjacent neighbors. In results presented subsequently, Student's *t* values for *I* and *c* are presented to indicate significant differences from zero. If forest volume is continuous, it is expected that these values will be positive and significant.

Table 12.1. Data Sets and Subsets Treated, Motivation for Subset, and Number of Points in the Validation and Calibration Subsets of Each

Data (Sub)set	Motivation	Points: Total	Cal	Val
1500 ha (between stand interpolability)				
All plots	Global interpolability	92	77	15
Softwood plots	Interpolability within species	81	69	12
Fir plots ≥60% density	Interpolability of fully stocked stands	61	52	9
16 ha (within stand interpolability)				
All plots	Global interpolability	81	67	14
≥ 250 stems per ha	Across species types w/o clear-cuts and blowdowns	65	54	11
Softwood and ≥250 stems per ha	Within a single (dominant) species type	55	46	9
Benchmarks				
Topographic	Known interpolable phenomenon	1037	998	39
Random	Noninterpolable phenomenon	100	81	19
Synthetic forest	Explanation of results	36	29	7

Second, a given data (sub)set was divided into a calibration and validation subset. The validation subset contained a random sample of interior plots that was approximately one-third of the total number of interior plots. The calibration data set contained the remaining plots. Spatial interpolation using Voronoi diagram area-stealing (Gold, 1989) was conducted on the calibration data set and the interpolated volumes for each of the validation data plot locations extracted. The mean difference between the true and interpolated values was calculated and tested for a significant difference from zero. The two values were also regressed against each other; perfect agreement will provide an R^2 of 1.0, a root-mean-square-error (RMSE) of 0.0, a slope of 1.0 and an intercept of 0.0.

The subsets of the two real-world forestry data sets examined were designed to test the interpolability of forest volume at a variety of levels under the conditions studied (Table 12.1). For the 92 plots over 1,500 ha, the subsets were (a) all data together, (b) only plots labeled as softwood on the forest map of the area, and (c) only plots labeled as predominantly fir with a density greater than 60%. For the 81 plots on a 16 ha grid, the subsets were (a) all data together, (b) only plots having 250 stems per ha and higher, and (c) only softwood plots having 250 stems per ha or higher.

To serve as a point of comparison, three additional data sets were also treated and analyzed in exactly the same manner as the other data (sub)sets. First, a digital terrain model for a 1-km (100 ha) square developed for another study was employed (Lowell, 1994b). Second, 100 locations were randomly generated over a 500-m square area and a value between 1 and 100 (with replacement) randomly assigned to each of the locations. These two data sets provide benchmarks for what one should find for a phenomenon known to be interpolable (elevation) and one which is not at all expected to be interpolable (random locations and attribute assignments). The third synthetic data set was a grid of 36 points spaced 50 m apart (Figure 12.1). The use of this data set will become clear after the results of the other data sets have been presented and discussed. For the moment, suffice it to say

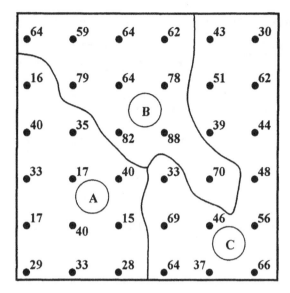

Figure 12.1. Synthetic forest stand. Ranges for types: (A) 10 to 40, (B) 60 to 90, (C) 30 to 70.

that this data set is designed to simulate three different forest types that were sampled systematically and densely. The value assigned at each location was random within the range assigned a priori to each type: (A) 10 to 40, (B) 30 to 70, and (C) 60 to 90. The overlap between class volumes is intentional as it reflects real forest conditions wherein it is possible that the volume at any given location for a particular type is less/more than that found in a type known to contain less/more volume. The mean, of course, of the two types will reflect their relative densities.

RESULTS AND DISCUSSION

In the following discussion, readers are reminded that the sample sizes are not the same for all data (sub)sets considered (see Table 12.1). In general, statistics such as R^2 and the statistical significance of various parameters cannot always be interpreted without consideration for sample size. However, in this study, calibration sample sizes are large enough in all cases to be directly comparable. Moreover, validation sample sizes are small enough that, if something appears to be statistically significant, this finding is not merely an artifact of a large sample size. In this study, therefore, differences in sample sizes do not affect overall conclusions and can be compared directly.

The results for the two benchmark data sets provide a useful manner in which to assess the results for the real-world forest data sets (Table 12.2). In general, both the topographic and the random data sets perform as expected.

The topographic data are strongly and positively spatially autocorrelated, suggesting that interpolation for this phenomenon at the scale sampled is warranted. This is further reinforced by noting the strong relationship that exists between the true and interpolated data for the validation data points ($R^2=0.97$), and the small mean difference between the two sets of values. Moreover, the slope and intercept of the regression line are close to the

Table 12.2. Results of Analysis for Spatial Autocorrelation and Interpolation

Mean[a] Data Set	Spat. Auto.[b]			Regression[c]				All Data[d]	
	Diff.	*I*	*c*	R^2	Slope	Intrcpt	RMSE	Mean	Std Dev
1500 ha (between stand interpolability)									
All plots	−18.6	0.98	0.62	0.16	0.95	−12.5	65.5	118.6	62.8
Softwood plots	36.5	0.97	0.93	0.00	−0.38*	139.1	70.1	116.8	63.8
Fir plots ≥60% density	−30.8	0.53	0.41	0.00	0.08**	90.0	62.1	121.3	62.2
16 ha (within stand interpolability)									
All plots	2.5	4.02**	3.56**	0.00	0.14	66.9	41.7	73.7	43.2
≥ 250 stems/ha	2.1	3.25**	1.88	0.00	0.38	54.0*	22.3	89.5	31.6
Softwood and ≥250 stems/ha	−11.9	1.77	0.81	0.00	−0.24	97.3**	28.2	92.6	31.7
Benchmarks									
Topographic	−0.1	52.26**	47.36**	0.97	1.04	2.4	2.5	60.1	14.1
Random	−5.9	−1.01	−0.38	0.00	0.18*	39.6	31.6	52.6	28.6
Synthetic forest	−1.3	3.16**	2.02*	0.07	1.22	−12.7	28.2	48.4	19.9

[a] Validation data only.

[b] *t* values for Moran's *I* and Geary's *c.* for all data. */** indicates significant difference from zero at 95%/99%.

[c] Validation plots only. */** indicates significant difference from expected values at 95%/99%.

[d] All data for purposes of comparison.

expected values of 1.0 and 0.0, respectively. Another measure of the usefulness of interpolation can be seen in the value of the RMSE (2.5) compared to that of the standard deviation (14.1) for all data. In the absence of the interpolated values, the best estimate for a given point's elevation will be 60.1 (the overall mean) with the error varying ± 28 m with 95% confidence (a 95% confidence limit using the standard deviation under an assumption of a normal distribution). The interpolation will provide an estimate at a given location that varies only ± 5 m with 95% confidence.

By contrast, the random data set indicates no spatial autocorrelation structure (nonsignificant *I* and *c*), suggesting that this is not a continuous phenomenon. And, indeed, the R^2 indicates that no correlation exists between the interpolated and true values for the validation data points. Furthermore, the slope of the line is significantly different from the expected value of 1.0 (though not significantly different from a slope of zero which indicates no relation). In the absence of a useful interpolation, the value estimated at a given point will be the mean (52.6) ± 56.1 (95% confidence), while the interpolation gives estimates ± 61.9 units (95% confidence). Clearly, the random values should not have been interpolated as indicated by the spatial autocorrelation statistics and the subsequent regression analysis. In fact, the consequences of having conducted an interpolation when it was not warranted has been an increase in the error on the resulting surface. In a practical sense, this means that the resulting interpolation surface is inferior to simply assigning the mean value everywhere.

It is also apparent that none of the situations explored using the 92 sample plots are interpolable. Perhaps the quickest way to see this is to note that none of the interpolations provided RMSEs that were noticeably different from the standard deviations of the data.

However, this is also abundantly clear in the low R^2s, the coefficients of the regression lines, and the lack of significance of either spatial autocorrelation coefficient. The importance of these findings is considerable.

These demonstrate that, in the conditions studied, between-stand interpolation is not warranted. This is perhaps not surprising in the case of global interpolability (All plots) and interpolability within species (Softwood plots)—particularly considering that Montmorency Forest is a managed forest on which normal harvesting activities have been ongoing. Thus the volume of the stands represented in these two cases may reflect more on past management practices than on the biological productivity of these areas—even though plots were all located in relatively mature forests. Note that these conclusions give little idea about what would be found in forests which have not been subjected to widespread natural or human disturbances. The findings for the 92 plots also indicate that one cannot interpolate even for fully stocked stands of a similar species combination which have not been subjected to recent disturbance. These stands (Fir plots ≥ 60% density) should have had time to manifest their biological potential and possibly show a gradient of productivity across the roughly 7 km by 3 km study area. It might be suggested that the reason that this was not the case was that the productivity of each stand is related to the underlying ecophysiography of a stand, which was not considered here. However, in another study which used the same 92 sample plots (Coulombe and Lowell, 1995) the relation of the productivity of each site to its underlying ecophysiography as estimated by map-based data (soils, slope, etc.) was examined; no significant relationships were found. Thus it would appear that over a large area, even with a sampling density much higher than is usually employed, interpolation of forest volume is not appropriate even for stands which have remained undisturbed for a long period of time.

At first glance, the results for the various subsets of the 81 sample plots might indicate that within-stand interpolation is possible—both spatial autocorrelation coefficients are positive and relatively large for All plots and plots having ≥ 250 stems per ha. However, the associated R^2s are zero (0.0) and the RMSEs are not largely different from the standard deviations. A visual examination of the data makes it very clear that the reason the spatial autocorrelation statistics are strong in the case of All plots is the presence of four adjacent plots being located in a clear-cut, and another five plots being located in a blowdown area. Simply stated, low volumes cluster in these areas—i.e., are positively spatially autocorrelated. When these are removed, spatial autocorrelation decreases, though it remains positive and strong. While this suggests that interpolation would be appropriate, the regression statistics suggest otherwise.

This can be explained by considering the synthetic surface in Figure 12.1 and its associated statistics in Table 12.2. This surface was constructed so that large values would have a tendency to cluster. That this was the case is apparent from the spatial autocorrelation coefficients for this surface (Table 12.2). Despite these values, the results of the regression analysis demonstrate that if one removes a given observation and attempts to estimate its value from surrounding observations, the precision of estimates is fairly low. That is, though similar values show tendencies to group, within each group there is no consistency of estimates. This suggests that there are identifiable areas—i.e., polygons—of high and low values, but within those areas, the variance is spatially random.

That this is the case in real forests is further reinforced by considering the Softwood and ≥ 250 stems per ha plots. In this data subset, spatial autocorrelation is weakened to the point where it is no longer statistically significant, and the regression statistics show that interpolation has not provided better estimates than would the mean and the standard deviation alone. Given that the case of Softwood and ≥ 250 stems per ha can be considered a single forest type, if one cannot observe a spatial tendency for wood volume within a stand, then one should not conduct within-stand interpolation.

Note that in the absence of within-stand spatial dependency, the optimal volume estimate for all locations within a stand would be the mean for the stand. Although producing an optimal surface was not the goal of this study, it remains that some practitioners may wish to do this. Kriging is an interpolation method that is presently in vogue (see, for example, Delfiner and Delhomme, 1973). In kriging parlance, this is known as having a "high" or "pure" nugget effect, and kriging automatically assigns the mean to a region whose data indicate a pure nugget effect. However, this only occurs for a given stand—a subset of the region—if the kriging procedure "knows" a priori which observations belong to which stands—something that will not be known if one expects a priori to be able to treat forest volume as a continuous phenomenon. Moreover, experiments using kriging were conducted on the synthetic data of Figure 12.1. These demonstrated that kriging will indicate a strong spatial dependence of observations on this surface, thereby reflecting the fact that high values tend to cluster. However, the kriging analysis was ill-equipped to recognize that this was caused by natural groupings and not because of spatial continuity over the entire region. Consequently, kriging produced estimates for validation points that were no better than those already presented.

To understand all of the findings of this study—i.e., a complete lack of interpolability—one must consider the nature of forest sampling. With conventional forest inventory techniques, the volume that is found at any given location is somewhat a function of the microlocation of the plot. That is, each forest type has an associated range of volume that one will find at any single location. When one samples a given location, one cannot know if one is at the upper, lower, or middle portion of that data range. Thus a large part of the reason that both between- and within-stand interpolability is inappropriate is simply the natural random spatial variability of a forest type and the way in which it is sampled. One could minimize such effects by using a larger sample unit so that one is consistently sampling the mean of a type (or at least that the range for the type would be reduced). But doing so risks requiring an excessive amount of field time and also requiring sample plots that exceed the boundaries of the forest stand being sampled. Subsequent data would therefore not be representative and would not accomplish what was intended.

It might be argued that the solution to this latter problem is to increase the size of the sample plots while decreasing the number of plots, so that the total field effort remains the same. However, little work has been conducted to determine the sample plot size necessary to produce a stable mean. Moreover, even if the optimal size were known, decreasing the number of sample plots below a certain level is not an option since a reliable estimate of population variance would no longer be available. It remains, therefore, that there are a number of statistical and operational barriers to developing alternative sampling units that might be more suitable for the work conducted.

Finally, it is noted that the forest studied represents a single set of forest conditions. In Montmorency Forest, these are naturally regenerated stands that have not had any silvicultural treatments during their lifetime. However, had the stands been manipulated by thinning, for example, it is anticipated that results would not be different than those presented, since volume would have been artificially controlled at some point in the life of the stand. However, in the forest conditions studied, it is possible that high natural forest variability masked all spatial dependence among volume. Consequently, plantation forests might allow a general trend to manifest itself—something that would not have occurred herein.

CONCLUSIONS

It does not appear to be appropriate to model the forest as a continuous surface. Instead, the results from this study suggest that the forest is a set of polygons. However, this does not mean that these are polygons with crisp boundaries as is usually represented on forest maps. While the synthetic surface employed herein (Figure 12.1) was based on definite, widthless boundaries, it is apparent that a number of different interpreters presented with only the 36 points in Figure 12.1 would not develop the same map. And even if such interpreters were provided with the knowledge that there were only "truly" three types present, it is doubtful that the lines would be placed in the same locations by different interpreters. Thus, the best interpretation of how to represent a forest might be as a set of polygons whose boundaries show a range of possible locations.

ACKNOWLEDGMENTS

The financial support for this work provided by the Association of Quebec Forest Industries and the Natural Sciences and Engineering Research Council of Canada within the Industrial Research Chair in Geomatics Applied to Forestry of Laval University is gratefully acknowledged.

REFERENCES

Beaulieu, P. and K. Lowell. Spatial autocorrelation among forest stand identified from the interpretation of aerial photographs, *Landscape and Urban Planning*, 29, pp. 161–169, 1994.

Bélanger, L., L. Bertrand, P. Bouliane, and L.-J. Lussier. *Plan d'aménagement de la forêt Montmorency*, Université Laval, Faculté de foresterie et de géomatique, Internal document, 1988.

Burrough, P. *Principles of Geographical Information Systems for Land Resources Assessment*, Clarendon Press, Oxford, 1986.

Coulombe, S. and K. Lowell. Ground-truth verification of relations between forest basal area and certain ecophysiographic factors using a geographic information system, *Landscape and Urban Planning*, 32, pp. 127–136, 1995.

Delfiner, P. and J. Delhomme. Optimum interpolation by Kriging, *Display and Analysis of Spatial Data*, John Wiley & Sons, 1973, pp. 96–114.

Gold, C.M. Surface interpolation, spatial adjacency, and GIS, Chapter 3, in *Three Dimensional Applications in Geographic Information Systems*, J. Raper, Ed., Taylor and Francis, London, 1989, pp. 21–35.

Griffith, D.A. *Spatial Autocorrelation: A Primer*, Monograph of the American Association of Geographers, Washington, 1985.

Joy, M. and M. Klinkenberg. Handling uncertainty in a spatial forest model integrated with GIS, in *GIS '94: Proceedings of the 8th Annual Symposium on Geographic Information Systems*, Vancouver, BC, 1994, pp. 359–365.

Lowell, K.E. Fuzzy cartographic representation for forestry based on Voronoi diagram area stealing, *Can. J. For. Res.* 24, pp. 1970–1980, 1994a.

Lowell, K.E. Probabilistic temporal GIS modelling involving more than two map classes, *Int. J. Geogr. Inf. Syst.* 8, pp. 73–93, 1994b.

Lowell, K.E. Effects of adjacent stand characteristics and boundary distance on density and volume of mapped land units in the boreal forest, *Plant Ecology* (recently renamed from *Vegetatio*), in press.

Chapter 13

Statistical Models of Landscape Pattern and the Effects of Coarse Spatial Resolution on Estimation of Area with Satellite Imagery

Christine A. Hlavka

INTRODUCTION

Changing land use/land cover raises important scientific questions about global climate, emissions of greenhouse gases, and energy exchange rates. These are being addressed by federal programs such as NASA's Mission to Planet Earth (MTPE) and by international efforts such as the International Geosphere-Biosphere Program (IGBP). For example, it is possible that increased wetland areas due to higher air temperatures will be a positive feedback on global warming, because the wetlands are a major source of methane (Aselman and Crutzen, 1989; Matthews and Fung, 1987), a greenhouse gas. The pervasive use of fires for land clearing and pasture management in the tropics makes biomass burning a significant source of carbon dioxide (Houghton et al., 1985; Detwiler and Hall, 1988) and other greenhouse gases (Crutzen and Andrae, 1990) that may be contributing to global warming.

The quantification of these effects depends on reliable estimates of global land use/land cover area (Aselman and Crutzen, 1989) but only rough estimates are available (Aselman and Crutzen, 1989; Matthews, 1983; Matthews and Fung, 1987; Robinson, 1989). To address the requirement for reliable estimates, methods for producing regional to global maps with satellite imagery are being developed. For example, the Brazilian Institute for Space Research (INPE) has developed a national map of weekly fire activity using Advanced Very High Resolution Radiometer (AVHRR) imagery (Setzer and Periera, 1991). Methods for developing maps of high latitude wetlands using European Remote-Sensing Satellite (ERS1/2) and Radarsat synthetic aperture radar (SAR) imagery are being developed (Morrissey et al., 1996).

The only practical way to produce maps over large regions of the globe is with data of coarse spatial resolution, such as AVHRR imagery at 1.1 km resolution or ERS-1 imagery at 240 m resolution. It would not be practical to use fine resolution satellite data, such as that provided by Landsat Multispectral Sensor (MSS, 80 m resolution), for mapping a large region because of the enormous volume of data that would be required. A single

AVHRR data set typically covers the same area as 100 Landsat scenes. The problem with using coarse resolution data is that the accuracy of the resultant maps is in doubt, especially in highly fragmented land cover types with a large proportion of border; i.e., mixed, pixels (Markham and Townshend, 1981; Woodcock and Strahler, 1987; Mayaux and Lambin, 1995). Errors on the order of 40–50% have been reported for estimates of the extent of burn scars (Setzer and Pereira, 1991) and forest classes (Moody and Woodcock, 1994).

Efforts to solve the problems associated with using coarse resolution imagery have involved regression of apparent area on coarse resolution classifications (Nelson et al., 1987; Mayaux and Lambin, 1995; Moody and Woodcock, 1996) or spectral values of coarse resolution imagery (Iverson et al., 1994) versus fine-scale data in sample areas. It has, however, proved difficult to acquire sufficient fine-scale data to develop the regression parameters. The relationship between area on coarse and fine-scale classifications depends on the degree of fragmentation of the map classes and that varies from site to site over the study area.

An approach using mathematical relationships among elements in the landscape, as observed with coarse imagery, might be the foundation for a modeling or self-calibrating type of procedure for estimating area with coarse resolution imagery. An example of such an approach is the "scale-up" estimator proposed by Maxim and Harrington (1982) for improving area estimates based on photo-interpreted aerial photography. The "scale-up" concept is illustrated by the example shown in Figure 13.1. Fragments of a cover type are accurately mapped down to some cut-off size C, and smaller fragments are not detected. The size distribution of the fragments in the scene is modeled by the function $D(a;A,B,...)$ for $a \geq 0$ and the observed size distribution is modeled by the censored distribution: $D(a;A,B,...)$ for $a \geq C$, and zero otherwise. For some functional forms, it is possible to estimate the parameters A,B,... and cut-off value C from the detected fragment sizes. The "scale-up" estimator is the total area of the detected fragments multiplied by the ratio of the total areas predicted by the models:

$$A_{scaled} = A_{observed} * \int_0^\infty D(a;A,B,...)da / \int_C^\infty D(a;A,B,...)da$$

If the number of fragments of size a is proportional to e^{-Ba}, this factor is $e^{BC}/(1 + BC)$. Note that this procedure does not require calibration data for implementation, but instead relies on a model for both the observed data and underlying scene characteristics.

The feasibility of the above modeling type of approach for estimating area has been explored with maps of burn scars in the savannas of Central Brazil (Hlavka et al., 1996) and maps of open water in Alaskan tundra (Morrissey et al., 1996) developed with satellite imagery. Burn scars and ponds are both highly fragmented types of land cover, and are therefore difficult to accurately map with coarse resolution imagery. They are suitable subjects for modeling with statistical distributions, due to the large number of fragments such as the ponds accounting for much of the surface water in Barrow, Alaska (Figure 13.2). This chapter presents the technical approach for characterizing size distributions from digital maps, assessing the effects of coarse resolution on size distributions, and relating the results to feasibility of a model-based method of area estimation.

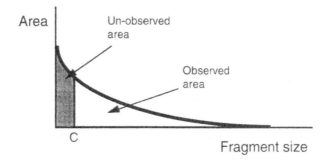

Figure 13.1. The "scale-up" concept. Total area of fragments is modeled as the area under a smooth parametric curve. In this example, the area is divided into two components—the observed and unobserved areas.

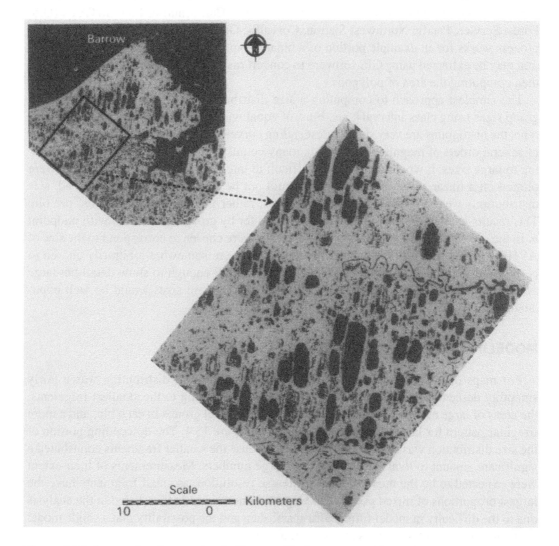

Figure 13.2. A map of surface waters in Barrow, Alaska generated from 12.5 meter ERS-1 radar imagery.

OBSERVING SIZE DISTRIBUTIONS

Size distributions of fragments were computed on both relatively fine-scale digital maps (processed Landsat MSS and ERS-1 12.5 meter imagery) and a coarse-scale map of burn scars derived from AVHRR imagery.

Each fragment in a digital map in raster format is a group of neighboring pixels coded as belonging to the target class. The observed size of each fragment is computed as the number of pixels in the group multiplied by pixel extent; i.e., the product of the distances between pixel centers in the vertical and horizontal directions. The number of pixels in each connected group is computed by recoding each pixel with a unique group code followed by computing the number of pixels per code number using ERDAS Imagine (Earth Resources Data Analysis Systems, Inc., Atlanta, Georgia). The pixel group sizes can also be computed in a similar manner with other image processing or GIS software such as ARC/INFO Grid (Environmental Systems Research Institute, Redlands, California) or FRAGSTATS (U.S. Forest Service, Pacific Northwest Station, Corvallis, Oregon). Figure 13.3 shows how this process works for an example portion of a binary map. Approximately the same information may be extracted using GIS software to convert raster formatted maps to polygons and then computing the area of polygons.

The simplest approach to computing a size distribution is to make histograms of the group sizes using class intervals; i.e., bins, of equal width. For a highly fragmented cover type, the histograms are very sharply descending curves with counts encompassing a range of several orders of magnitude, including many counts of zero or one in bins corresponding to large sizes. It would therefore be difficult to interpret such histograms if sizes were plotted on a linear scale, and a log scale could not be used ($\log(0) = -\infty$). Instead, size distributions were computed by summing sizes, rather than computing counts, per bin. This results in tabulations of area $D(a)$ accounted for by groups in each bin with midpoint a. In the case of the burn scar maps, the bin widths were chosen to correspond to the size of AVHRR pixels. For the water map, the bin widths were somewhat arbitrarily chosen to correspond to the area of 20 pixels (0.0034 km^2)—small enough to show detail but large enough so that the bins corresponding to the small fragment sizes would be well populated.

MODELING SIZE DISTRIBUTIONS

For maps developed from the finer-scale imagery, the size distribution was a fairly smoothly descending curve for a range of sizes corresponding to the smallest fragments, the areas of large numbers (hundreds to thousands) of pixel groups in each bin, and a more irregular pattern for larger fragments, as shown in Figure 13.4. The descending portion of the size distribution was selected for modeling because the smaller fragments contributed a significant amount to total area, due to their large numbers. Measurements of their extent were expected to be the most affected by coarse resolution, as small fragments have the largest proportions of mixed pixels. The larger fragments were not included in the analysis due to the difficulty in model fitting with sparse data and the possibility that a single model might not be appropriate across the entire range of sizes.

Two types of distributions that describe predominance of small targets in overall area were fitted to the size distribution. The power, or Pareto distribution (Johnson and Kotz,

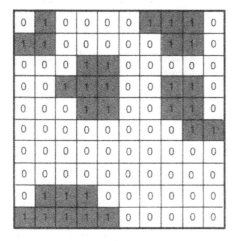 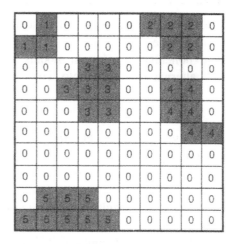

binary map   recoded map

Figure 13.3. Recoding of binary maps to label individual groups. (a) Binary map input, (b) Recoded map.

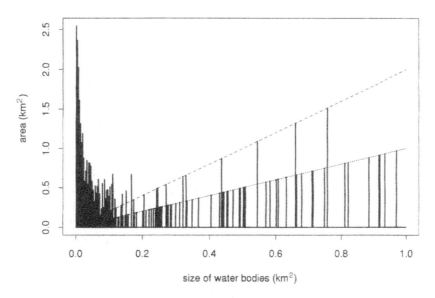

size of water bodies (km²)

Figure 13.4. The size distribution D(*a*) for water bodies under 1 km² in the map for Barrow, Alaska (Figure 13.2). The bin midpoints *a* were 0.00172 km², 0.00484 km², 0.00796 km², 0.0111 km², ... Note the spike of large values in the first few bins due to many small water bodies and the sparse distribution for sizes greater than 0.1 km², with zero, one (bar tops following the dashed 45° line) or two occurrences per bin (following the dashed 63° line).

1970), has been related to fractal patterns in the landscape (Mandelbrot, 1983) and other fractal phenomena (Harrison et al., 1978). The power distribution:

$$D_1(a) = K_1 a^{-B}$$

was fitted by regressing $\log[D(a)]$ versus $\log(a)$ to estimate K_1 (regression intercept) and B (regression slope).

The exponential distribution models less steeply descending curves. It therefore is a practical alternative to the Pareto. The exponential distribution:

$$D_2(a) = K_2\, e^{-Ga}$$

was fitted by regressing $\log[D(a)]$ versus a to estimate K_2 (regression intercept) and G (regression slope).

Figure 13.5 illustrates the model fit for small water bodies. For size distributions of the small targets in binary maps tested (two Landsat MSS maps of burn scars, one ERS-1 map of open water), the correlation of predictions of one or the other of the two models with observed points was better than 90%. In the case of the Landsat burn scar maps, the exponential model fit best (R values of 99% and 90%), while the power model fit the water body data with R = 97%.

EFFECTS OF COARSE RESOLUTION

A map of burn scars generated from processing an AVHRR image was used for a comparison with one of the Landsat maps. The satellite overpasses were less than one day apart and the visual quality of the images was good, with little haze obscuring the ground. The contrast between scars and other land cover was good, particularly in the near infrared bands (Landsat MSS4 and AVHRR2). Differences in the characteristics of the size distributions of small scars can therefore be attributed to the effects of spatial resolution of the sensors.

Some effects of coarse resolution are illustrated in Figures 13.6 and 13.7. The omission effect, as described by Maxim and Harrington (1982), decreased the value of $D_{AVHRR}(a)$ for the first point (bin) on the left. The value of the cut-off was estimated to be about a seventh of a pixel (Hlavka and Livingston, 1997). In addition, there was a second effect due to the characteristics of coarse digital satellite imagery. The second, third, and fourth values for $D_{AVHRR}(a)$ are larger than corresponding values of $D_{MSS}(a)$, due to small scars overlapping neighboring pixels. To illustrate this effect, Figure 13.7 shows a scar about the size of one pixel overlapping, and being detected, in a pair of neighboring pixels. This "overlap commission error" is exaggerated by the overlap of the fields of view (FOVs) associated with the pixels. In the case of AVHRR, the FOV, or sensor "footprint," is about 40% broader than the center-to-center pixel spacing (Kidwell, 1991). The difference between the areas under the curves in Figure 13.6 was about 30%, indicating overestimation due to coarse resolution in this case.

A third effect of coarse resolution is that individual fragments may not be resolved in the imagery. This effect was not seen in the smaller burn scars in the AVHRR, but is a potential factor when fragment densities are generally or locally high. This "clumping effect" (Figure 13.7) will cause depression in the size distribution for small sizes and inflation for larger sizes, and increase the apparent total area.

CONCLUSION AND DISCUSSION

A modeling approach to estimating area with satellite imagery of coarse spatial resolution is of interest because implementation may require less fine-scale data than current

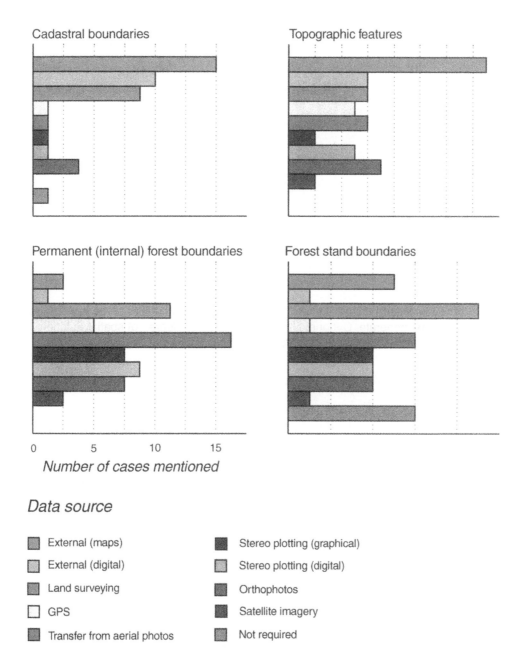

Plate 1. Sources of spatial data used by 25 forest management agencies.

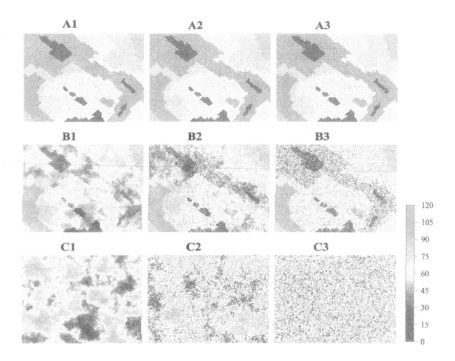

Plate 2. Nine simulated realities. Letter indicates the degree of influence by the artificial soil map: A=strong, B=moderate, C=none. Number indicates the degree of spatial autocorrelation in the added residual: 1=strong, 2=moderate, 3=none.

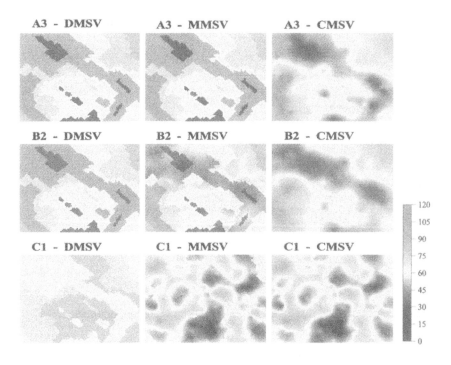

Plate 3. Mapping three simulated realities (A3, B2, and C1) from 200 observations using the DMSV, MMSV, and CMSV.

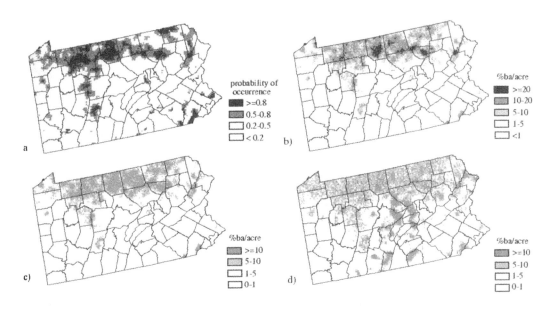

Plate 4. Four modeled data sets describing the distribution of beech in Pennsylvania: (a) the estimated probability of occurrence using indicator kriging, (b) the 70th percentile from 100 sgCS realizations, (c) the minus variation (70th–17th percentile), and (d) the plus variation (83rd–70th percentile) about the 70th percentile estimate.

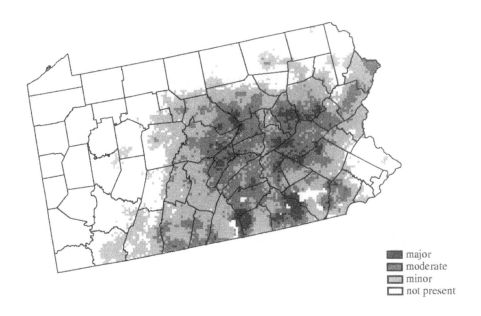

Plate 5. Chestnut oak occurrence as a major (>40% ba/acre), moderate (20–40%), and minor component (<20%) of the total ba/acre. Derived from sgCS estimates using the 75th percentile.

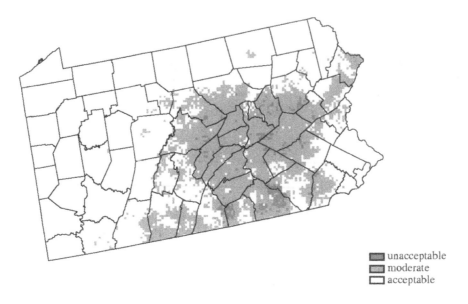

Plate 6. The uncertainty associated with the estimates used in Plate 5, in classes of: acceptable (≤–10% ba/acre), moderate (–10–25% ba/acre), and unacceptable (>–25% ba/acre). Derived from the sgCS estimates.

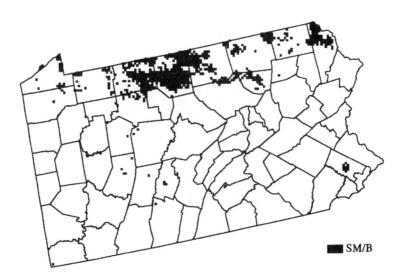

Plate 7. The occurrence of SM/B type, defined as those areas where the combined % ba/acre of sugar maple and beech amounts to more than 50% of the total ba/acre in the stand, and derived from the 70th percentile estimate from sgCS.

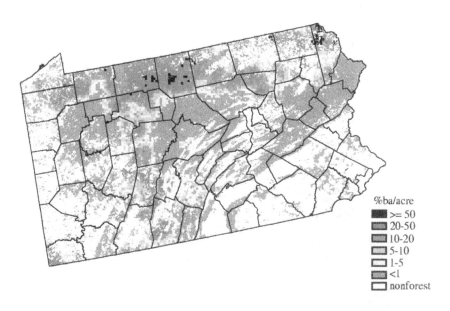

Plate 8. A "map" of sugar maple % ba/acre with nonforest areas masked out. The data set used is the 70th percentile of sgCS, and the nonforest area is derived from AVHRR data (Evans and Zhu, 1992).

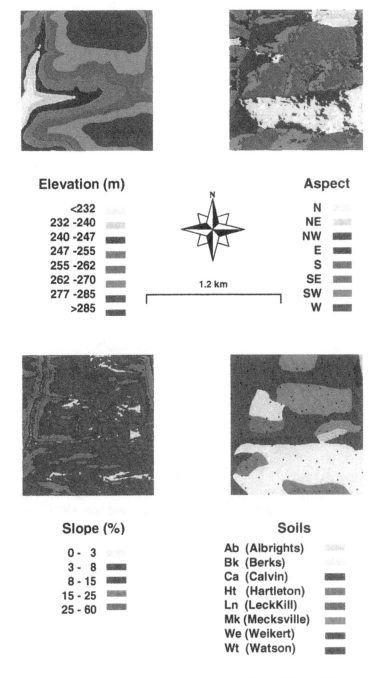

Elevation (m)

<232
232 -240
240 -247
247 -255
255 -262
262 -270
277 -285
>285

Aspect

N
NE
NW
E
S
SE
SW
W

1.2 km

Slope (%)

0 - 3
3 - 8
8 - 15
15 - 25
25 - 60

Soils

Ab (Albrights)
Bk (Berks)
Ca (Calvin)
Ht (Hartleton)
Ln (LeckKill)
Mk (Mecksville)
We (Weikert)
Wt (Watson)

Plate 9. Case Study Area: Maps of Elevation (a), Aspect (b), and Slope (c), based on a detailed 5-m digital elevation model (*DEM*); and the intersection of the mask of Soil Types (d) with the assigned locations of soil pedons.

SOIL WATER CONTENT (by vol)

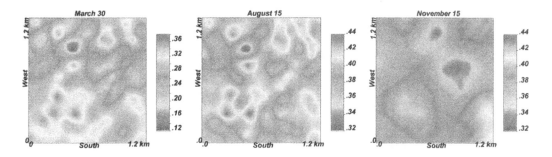

Plate 10. Kriged distributions $(OK)^{1}$ of soil moisture by volume θ_v in the surface (01), fourth (04), and sixth (06) soil layer at the Case Study site on March 30th, August 15th, and November 15th.

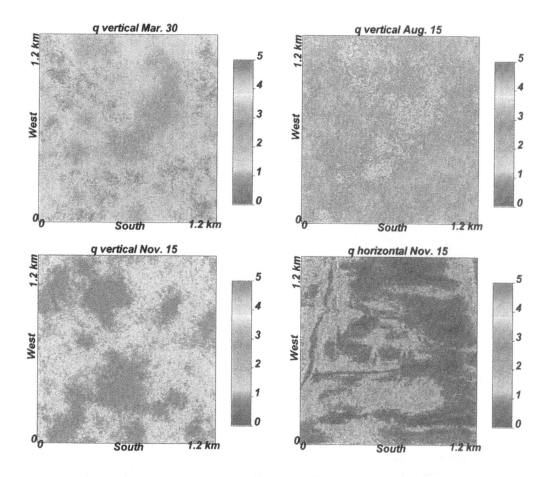

Plate 11. Conditionally simulated spatial distribution of expected values of recharge modeled as the vertical percolate flux q_v (cm) on March 30th, August 15th, and November 15th and for the interflow modeled as horizontal flux q_h (cm) on November 15th, based on 25 sequential indicator simulations (*sisim*) with 101 conditioning data.

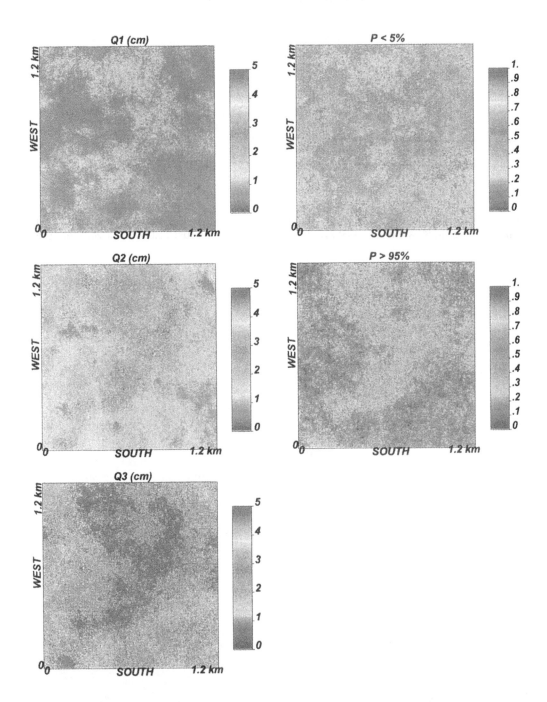

Plate 12. Conditionally simulated (*sisim*) vertical percolate flux q_v (cm) for the three quartiles (Q_1, Q_2, and Q_3) of the distribution on August 15th and the spatial distribution of the probability (P) of values of q_v being lower than Q_{05} (1.52 cm) or higher than Q_{95} (3.72 cm).

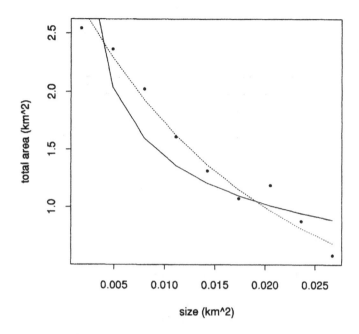

Figure 13.5. An example fit of the size distribution of ponds: the first nine points of Figure 13.4 to the power distribution (solid line) and exponential distribution (dotted line).

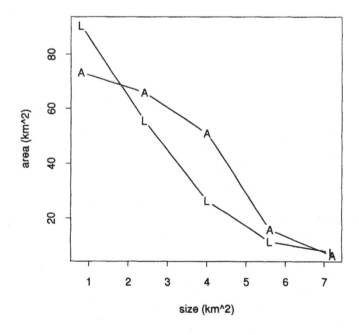

Figure 13.6. Size distribution of burn scars under 8 km² in and around Emas National Park, Brazil as mapped with Landsat (L) and AVHRR (A).

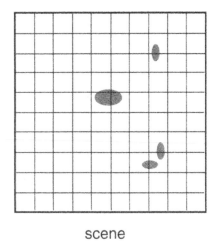 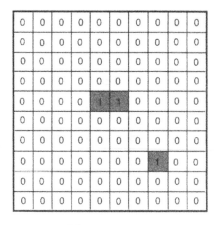

scene   binary map

Figure 13.7. Effects of coarse spatial resolution on the detection of fragments in a digital image. (a) Scene with overlay showing grid defined by image pixelation. (b) Binary map developed from digital imagery when detection occurs if overlap of target in a pixel is greater than a quarter of a pixel. Effects include omission of the small fragment in the right upper part of the scene, "overlap commission" of the fragment in the middle, and the "clumping" of the pair of small fragments in the lower right.

techniques based on empirical relationships between observations at fine and coarse resolutions. This approach, first proposed for improving the accuracy of area estimates based on aerial photography (Maxim and Harrington, 1982), is based on modeling the underlying and observed size distributions of the small patches of a highly fragmented cover type. Such an approach may be practical because software for computing the sizes of fragments in a digital map is widely available.

The size distributions of the smaller fragments on several relatively fine resolution digital maps of burn scars and open water have been examined to assess the feasibility of modeling the underlying size distributions. The size distributions associated with the small fragments in several example maps fit simple two parameter models with R values better than 90%. It therefore appears appropriate to model the numerous smaller fragments in a scene with a simple type of distribution.

The size distributions of small burn scars in an area of tropical savanna on maps developed from Landsat MSS and relatively coarse AVHRR imagery were compared. The effects of omission of the smallest burn scars and errors of commission associated with slightly larger scars were noted on the AVHRR size distribution, and accounted for a net overestimate in burn area by 30%. A model that accurately predicts observed area will therefore have to account for both omission and commission errors, and would not be as easy to construct as the proposed "scale-up" estimator (Maxim and Harrington, 1982) that only corrects for omission.

A simple size-dependent approach to improving area estimates for fragmented cover types may be effective if neighboring small fragments are resolved in the coarse imagery, and not "clumped" into larger patches. In this case, the effects of coarse resolution primarily affect fragments up to 5 pixels in extent. An estimate would be of the form:

$$A_{corrected} = F_c \sum_{a=1}^{4} D(a) + \sum_{a>4} D(a)$$

where F_c is a correction factor for the total area of the smaller fragments, and $D(a)$ is the total area of fragments with a pixels. The parameter F_c may be more robust than a correction factor applied to the total apparent area. The F_c term can be determined with fine-scale information on a relatively small set of sample areas, given the ubiquity of small fragments. Alternatively, it may be possible to devise a procedure based on modeling the underlying and observed size distributions by using knowledge of the functional form of the underlying distribution and the effects of coarse resolution. Such a procedure may require little or no fine-scale data.

ACKNOWLEDGMENTS

Technical advice on the coregistration of MSS and AVHRR imagery was provided by Richard Lucas of the University College of Swansea. The maps of burn scars developed from Landsat imagery were provided by Vince Ambrosia (Johnson Control World Services, NASA/Ames Research Center). The map of surface waters in Barrow, Alaska developed from ERS-1 SAR imagery was provided by Gerry Livingston and Leslie Morrissey (both of the University of Vermont), and Joel Stearn (Johnson Controls World Services, NASA/Ames Research Center). Funding for research on biomass burning and Arctic wetlands was provided by separate grants from the Terrestrial Ecology Program, part of NASA's Mission to Planet Earth. Jennifer Dungan (Johnson Control World Services, NASA/Ames Research Center) provided advice and support for the graphic presentations in this chapter. The author thanks two anonymous reviewers for suggestions that guided the revision of this chapter.

REFERENCES

Aselman, I. and P.J. Crutzen. Global Distribution of Natural Freshwater Wetlands and Rice Paddies, Their Net Primary Productivity, and Possible Methane Emissions. *J. Atmos. Chem.* 8, pp. 307–358, 1989.

Crutzen, P.J. and M.O. Andrae. Biomass burning in the tropics: Impact on atmospheric chemistry and biogeochemical cycles, *Science,* 250, pp. 1670–1678, 1990.

Detwiler, R.P. and C.A.S. Hall. Tropical forests and the global carbon cycle, *Science,* 239, pp. 42–47, 1988.

Harrison, R.J., G.H. Bishop, Jr., and G.D. Quinn. Spanning length of percolation clusters, *J. Stat. Phys.* 19, pp. 53–64, 1978.

Hlavka, C.A., V.G. Ambrosia, J.A. Brass, and A.R. Rezendez. Mapping Fire Scars in the Brazilian Cerrado Using AVHRR Imagery, in *Biomass Burning and Global Change,* J.S. Levine, Ed., MIT Press, Cambridge, MA, 1996.

Hlavka, C.A. and G.P. Livingston. Statistical models of fragmented land cover and the effects of coarse spatial resolution on the estimation of area with satellite sensor imagery, *Int. J. Remote Sensing*, pp. 2253–2259, 1997.

Houghton, R.A., R. Boone, J.M. Mellilo, C.A. Palm, G.M. Woodwell, N. Myers, B. Moore III, and D.L. Skole. Net flux of carbon dioxide from tropical forests in 1980, *Nature,* 316, pp. 617–620, 1985.

Iverson, L.R., E.A. Cook, and R.L. Graham. Regional Forest Cover Estimation via Remote Sensing: the Calibration Center Concept, *Landscape Ecol.* 9, pp. 159–174, 1994.

Johnson, N.L. and S. Kotz. *Continuous Univariate Distributions-1*, John Wiley & Sons, New York, 1970, pp. 233–247.

Kidwell, K.B. *NOAA Polar Orbiter Data (TIROS-N, NOAA-6, NOAA-7, NOAA-8, NOAA-10, and NOAA-12) Users Guide*, Washington, DC, NOAA/NESDIS, 1991.

Mandelbrot, B.B. *The Fractal Geometry of Nature*, W.H. Freeman and Co., New York, 1983, pp. 117–119.

Markham, B.L. and J.R.G. Townshend. Land Cover Classification Accuracy as a Function of Sensor Spatial Resolution, in *Proceedings of the 15th International Symposium on Remote Sensing of Environment, Ann Arbor, Michigan, 1991*. Environmental Research Institute of Michigan, Ann Arbor, 1981, pp. 1075–1090.

Matthews, E. Global vegetation and land use: New high resolution data bases for climate studies, *J. Climatic Appl. Meteorol.* 22, pp. 474–487, 1983.

Matthews, E. and I. Fung. Methane emissions from natural wetlands: Global distribution, area, and environmental characteristics of sources, *Global Biogeochem. Cycles,* 1, pp. 61–86, 1987.

Maxim, D.L. and L. Harrington. "Scale-up" estimators for aerial surveys with size-dependent detection, *Photogrammetric Eng. Remote Sensing,* 45, pp. 1271–1287, 1982.

Mayaux, P. and E.F. Lambin. Estimation of tropical forest area from coarse spatial resolution data: A two-step correction function for proportional errors due to spatial aggregation, *Remote Sensing Environ.* 53, pp. 1–15, 1995.

Moody, A. and C.E. Woodcock. Scale-dependent errors in the estimation of land-cover proportions: Implication for global land-cover datasets, *Photogrammetric Eng. Remote Sensing*, 60, pp. 585–594, 1994.

Moody, A. and C.E. Woodcock. A Calibration-Based Model for Correcting Area Estimates derived from Coarse Resolution Land Cover Data, Presented at the *Second International Symposium on Spatial Accuracy Assessment in Natural Resources and Environmental Sciences, May 21–23, 1996*, Fort Collins, CO, 1996.

Morrissey, L.A., S.L. Durden, G.P. Livingston, J.A. Stearn, and L.S. Guild. Differentiating methane source areas in arctic environments with multitemporal ERS-1 SAR data, *IEEE Trans. Geosci. Remote Sensing,* 34, pp. 667–673, 1996.

Nelson, R., N. Horning, and T.A. Stone. Determining the rate of forest conversion in Mato Grosso, Brazil, using Landsat MSS and AVHRR data, *Int. J. Remote Sensing,* 8, pp. 1767–1784, 1987.

Robinson, J. M. On the uncertainties in the computation of global emissions from biomass burning. *Climate Change.* 14, pp. 243–262, 1989.

Setzer, A.W. and M.C. Periera. Operational Detection of Fires in Brazil with NOAA-AVHRR, presented at the 24th International Symposium on Remote Sensing of the Environment, Rio de Janeiro, Brazil, 27–31 May, 1991.

Woodcock, C.E. and A.H. Strahler. The factor of scale in remote sensing, *Remote Sensing Environ.* 21, pp. 311–332, 1987.

Chapter 14

The Simultaneous Nature of Tree Growth Models

Hubert Hasenauer, Robert A. Monserud, and Timothy G. Gregoire

INTRODUCTION

Single-tree stand simulators (e.g., Monserud, 1975; Wykoff et al., 1982; Burkhart et al., 1987; Hasenauer, 1994) usually consist of different equations for predicting periodic diameter or basal area increment, height increment, and the probability of mortality for each sample tree. These equations commonly are assumed to be independent, with parameters of each equation estimated separately rather then simultaneously with linear or nonlinear regression (Hasenauer et al., 1998).

From a biological standpoint, it seems reasonable to assert that the change in a tree's basal area, height, and risk of mortality are not uncorrelated phenomena (Dixon et al., 1990). Depending on their interrelationships, simultaneous estimation of the models' parameters may be necessary in order to provide estimates that are consistent, or it may be desirable in order to provide more precise estimates than can be obtained otherwise

Seminal work by Aitken (1934–35), Haavelmo (1943), Theil (1953), Zellner (1962), and Zellner and Theil (1962) resulted in almost all methods currently available for estimating the parameters in intercorrelated (simultaneous) systems of equations. The advantage of simultaneous regression procedures is that the joint covariance matrix is estimated without implicitly restricting the between-equation covariances to be zero. Simultaneous regression techniques will lead to more efficient estimators if either (1) endogenous variables appear on the right-hand side, or (2) the contemporaneous correlations among the stochastic error terms are nonzero, or both. A gain in efficiency will increase the precision of the resulting model predictions because simultaneous estimation procedures are designed to be robust to departures from the true correlation pattern. Although we do not know the true correlation, those correlation models closer to the true correlation structure will realize greater efficiency because additional information is used to describe the system. In general, the gain in efficiency is higher when the errors among different equations are highly correlated (Judge et al., 1980).

The objective of this chapter is to explore the simultaneous nature of tree growth models. We estimate the parameters of individual tree models for basal area increment, height increment, and crown ratio using least squares methods separately and simultaneously by applying two-stage (Theil, 1953) and three-stage least squares (Zellner and Theil, 1962) techniques. We investigate simultaneous versus independent regression techniques by evaluating the cross-equation correlation, a diagnostic tool for determining the strength of the correlations among the predictions (Hasenauer et al., 1998).

METHODS

Independent Regressions

We begin with a system of three individual tree growth equations for stand conditions in Austria: basal area increment, height increment, and crown ratio. These equations were developed independently using the same data set.

Basal Area Increment

After eliminating the qualitative site descriptors chosen by Monserud and Sterba (1996), which reduced the variance explained only by 2.6%, we are left with the following model for Y_1:

$$
\begin{aligned}
\ln(\Delta BA) = a + b_1 \cdot \ln(D) + b_2 \cdot D^2 + b_3 \cdot \ln(C) \\
+ c_1 \cdot BAL + c_2 \cdot CCF^2 + d_1 \cdot ELEV^2 + d_2 \cdot SL^2 + \epsilon_1
\end{aligned} \tag{1}
$$

with BA the 5-year basal area increment (outside bark), D the diameter at breast height (1.3 m) in cm, $C=(1/CR)-1$ where CR is the crown ratio, BAL the basal area (m²/ha) of trees larger in diameter than the subject tree, CCF the crown competition factor of Krajicek et al. (1961), $ELEV$ the elevation in hectometers, and SL the tangent of the slope angle (%/100).

Height Increment

Hasenauer and Monserud (1997) used a similar formulation to predict 5-year height increment H, where H is the tree height, and all other parameters as previously defined. The second equation (Y_2) in the system is:

$$
\begin{aligned}
\ln(\Delta H) = a + b_1 \cdot \ln(D) + b_2 \cdot H^2 + b_3 \cdot \ln(C) \\
+ c_1 \cdot BAL + c_2 \cdot CCF + d_1 \cdot ELEV^2 + d_2 \cdot SL + \epsilon_2
\end{aligned} \tag{2}
$$

Crown Ratio

To ensure that the predictions of crown ratio (defined as the crown length divided by tree height) are bounded between 0 and 1, Hasenauer and Monserud (1996) chose a logistic function. After linearizing the logistic and rearranging, we are left with the following logarithmic transformation of crown ratio $(Y_3=ln[C])$:

$$
\begin{aligned}
\ln(C) = a + b_1 \cdot (H/D) + b_2 \cdot H + b_3 \cdot D^2 + c_1 \cdot BAL + c_2 \ln(CCF) \\
+ d_1 \cdot ELEV + d_2 \cdot SL^2 + d_3 \cdot SL \cdot \cos AZ + d_4 \cdot SL \cdot \sin AZ + \epsilon_3
\end{aligned} \tag{3}
$$

where $C=(1/CR)-1$, H/D is the height/diameter ratio (m/cm), AZ is the azimuth in radians, and all other parameters are as previously defined. Change in crown ratio is not available, because height to crown base was not remeasured after the initial inventory.

Simultaneous Equation Systems

In the system above, it is likely that the errors $\in$ in Equations 1–3 are intercorrelated because they are associated with various attributes of the same tree. If this is the only common influence among the three equations, then Zellner's (1962) seemingly unrelated regression (SUR) procedure would be appropriate because the equations are related through contemporaneous correlations in the variance-covariance matrix. We write the multivariate regression model as

$$Y = X\beta + \in \tag{4}$$

where Y is a $3n \times 1$ vector of dependent (endogenous) variables, X is the $3n \times (p_1+p_2+p_3)$ design matrix, β is the $(p_1+p_2+p_3) \times 1$ vector of coefficients to be estimated, and $\in$ is the $3n \times 1$ error vector. The errors $\in$ have fixed mean $E[\in]=0$ and variance

$$V[\in] = E[\in \in'] = \begin{pmatrix} W_1 & W_{12} & W_{13} \\ W_{21} & W_2 & W_{23} \\ W_{31} & W_{32} & W_3 \end{pmatrix} = \Omega \tag{5}$$

where $W_i=\sigma_i^2 I$ are the main diagonal variances and $W_{ij}=\sigma_{ij} I$ are the covariances, with I the n-dimensional identity matrix.

The ordinary least squares (OLS) estimator of $\beta=(\beta_1', \beta_2', \beta_3')$ is

$$b = (X'X)^{-1} X'Y \tag{6}$$

Under the conditions that X is entirely exogenous, all $\sigma_{ij}=0$ and all σ_i^2 are identically equal to some constant σ^2, then Equation 6 is an unbiased estimator of β and its variance is given by the familiar expression $(X'X)^{-1}\sigma^2$. For the more general error structure specified by Equation 5, the variance of b is

$$V(b) = (X'X)^{-1} X'\Omega X (X'X)^{-1} \tag{7}$$

If $\hat{\Omega}$ denotes a consistent estimator of Ω, then

$$\hat{V}(b) = (X'X)^{-1} X'\hat{\Omega} X (X'X)^{-1} \tag{8}$$

is a consistent estimator of $V(b)$. The restrictive conditions of the OLS estimator in Equation 6 assume that the between-equation covariances W_{ij} (see Equation 5) are zero. Thus $\hat{V}(b)$ collapses to $\hat{V}^*(b)$, where

$$\hat{V}^*(b) = (X'X)^{-1} \hat{\sigma}^2 \tag{9}$$

Failing these conditions, Equation 9 is neither an unbiased nor consistent estimator of $V(b)$.

In our system of growth equations, it is likely that the errors ε in Equations 1–3 are intercorrelated, for they are associated with various attributes of the same trees. If this intercorrelation were the only common influence among the three equations, then Zellner's (1962) seemingly unrelated regression (SUR) procedure would be appropriate because the equations are related through contemporaneous correlations in the variance-covariance matrix. However, the use of $ln(C)=Y_3$ on the right-hand side in the models for $ln(BAI)$ and $ln(\Delta H)$, Equations 1 and 2, respectively, precludes the straightforward use of SUR. Whenever a response or endogenous variable from one model appears as a predictor variable in another model, both the OLS and SUR estimator of β will be biased and inconsistent due to this stochastic regressor (Judge et al., 1980).

Although Theil's (1953) two-stage least squares (2SLS) can be used with stochastic regressors to provide consistent estimates of β, an even more efficient estimator was invented by Zellner and Theil (1962): three-stage least square (3SLS). By combining Theil's (1953) 2SLS procedure with Zellner's (1962) SUR procedure, the resulting 3SLS estimators of β are consistent and asymptotically more efficient than 2SLS. Zellner and Theil's (1962) 3SLS regression procedure is a form of feasible generalized least squares (Judge et al., 1980) in which Ω is estimated by $\tilde{\Omega}$.

The basic idea of 2SLS (Johnston, 1984) is to replace each right-hand side (RHS) endogenous variable with a nonstochastic estimate obtained by regressing the endogenous RHS variable on all the exogenous (fixed) variables in the model (Stage 1). This procedure replaces a stochastic predictor with a desirable instrumental variable that is contemporaneously uncorrelated with the disturbance and yet is nevertheless correlated with the RHS endogenous variable (Goldberger, 1964). In Stage 2, least squares regression is used with this nonstochastic instrumental variable along with the exogenous variables to estimate the parameters in a given model. In 3SLS, the variance-covariance matrix $\tilde{\Omega}$ is estimated from the residual covariance matrix of the 2SLS procedure and then the entire system is estimated simultaneously using generalized least squares (Stage 3). This retains the consistency property of 2SLS while increasing the asymptotic efficiency of the 3SLS estimator of β:

$$\tilde{\beta} = (X'\tilde{\Omega}^{-1}X)^{-1} X'\tilde{\Omega}^{-1}Y \tag{10}$$

The variance of $\hat{\beta}$ is estimated by

$$\tilde{\Sigma} = V(\tilde{\beta}) = (X'\tilde{\Omega}^{-1}X)^{-1} = \begin{pmatrix} \tilde{C}_{11} & \tilde{C}_{12} & \tilde{C}_{13} \\ \tilde{C}'_{12} & \tilde{C}_{22} & \tilde{C}_{23} \\ \tilde{C}'_{13} & \tilde{C}'_{23} & \tilde{C}_{33} \end{pmatrix} \tag{11}$$

Thus, 3SLS is consistent and asymptotically efficient when the system equations are related both through the disturbance terms and through RHS endogenous variables (Kmenta, 1986).

For our system of three growth equations, the estimate of the variance-covariance matrix consists of nine matrix partitions bold $\tilde{C}_{ij}$ with the following dimensions and attributes: $\tilde{C}_{11}$ is a 8×8 variance-covariance matrix for $\tilde{\beta}_1$, the logarithm of the predicted basal area increments. The main diagonal elements are the estimated variances, and the off-diagonal elements are the estimated covariances for all variables in the first equation. Similarly, $\tilde{C}_{22}$ and $\tilde{C}_{33}$ are 8×8 and 10×10 variance-covariance matrices for $\tilde{\beta}_2$ and $\tilde{\beta}_3$, respectively. The remaining off-diagonal partitions contain cross-equation covariances. For example, $\tilde{C}_{12}$ is an 8×8 covariance matrix containing the cross-equation covariances between each estimated parameter in the first growth equation (logarithm of basal area increment) and each estimated parameter in the second growth equation (logarithm of height increment). Because there are $p_1 + p_2 + p_3 = 8 + 8 + 10 = 26$ estimated parameters in the three growth equations, there are 676 elements in the symmetric variance-covariance matrix $\tilde{\Sigma}$, 351 of which are distinct. For further details concerning 3SLS procedures in a forestry context we refer to Borders and Bailey (1986) and Gregoire (1987).

In an individual tree model, estimates of $E[Y_1]$, $E[Y_2]$, and $E[Y_3]$ are needed for each tree k. Let x'_{1k} denote the $1 \times p_1$ row vector of covariate values for a particular tree k. The prediction for the basal area increment model for tree k is

$$\tilde{y}_{1k} = x'_{1k} \tilde{\beta}_1 \qquad (12)$$

which serves as an estimate of $E[y_{1k}|x'_{1k}]$. Let $\tilde{y}_{2k}$ and $\tilde{y}_{3k}$, the estimates for the height increment and crown ratio models for tree k be similarly defined. Note that x'_{1k} could be any row vector of covariates from any data set. In our case the cross-equation correlation will be evaluated on the same data set as it is used for predicting the estimates of $\tilde{\beta}_1$, $\tilde{\beta}_2$, and $\tilde{\beta}_3$.

The distributional properties of the random errors ε in conjunction with the estimator of β_i determine the statistical properties of the random variables $\tilde{\beta}_1$ and $\tilde{y}_{1k}$. For tree k, the scalar covariance between random variables $\tilde{y}_{ik}$ and $\tilde{y}_{jk}$ is $x'_{ik}\tilde{C}_{ij}x_{jk}$, where $\tilde{C}_{ij}$ is the $p_i \times p_j$ covariance matrix of $\tilde{\beta}_i$ and $\tilde{\beta}_j$. Therefore, the correlation $\tilde{r}_{ij,k}$ between $\tilde{y}_{ik}$ and $\tilde{y}_{jk}$ is

$$\tilde{r}_{ij,k} = \frac{x'_{ik}\tilde{C}_{ij}x_{jk}}{\sqrt{\left(x'_{ik}\tilde{C}_{ii}x_{ik}\right)\left(x'_{jk}\tilde{C}_{jj}x_{jk}\right)}} \qquad (13)$$

DATA

Data were obtained from the Austrian National Forest Inventory (Forstliche Bundesversuchsanstalt, 1981), a systematic hidden permanent sample plot design over the whole of Austria, with a 5-yr remeasurement interval. In a given year a fifth of the plots are remeasured, ensuring a representative sample of all Austrian forests each year. The total inventory comprises about 5,090 permanent plots in a single ownership and not crossed by roads, and includes more than 42,000 trees. We restricted ourselves to the 4,135 forested plots containing remeasured Norway spruce (*Picea abies* L. Karst).

Permanent sample plots were established from 1981 to 1985. Trees with a diameter at breast height (D, 1.3 m) larger than 10.4 cm were selected by angle count sampling (Bitterlich, 1948) using a basal area factor (BAF) of 4 m²/ha. Trees with a D between 5 and 10.4 cm were measured within a circle of 2.6 m radius located at plot center; smaller trees were not recorded. At plot establishment, the following data were recorded for every sample tree: species, D to the nearest mm, and distance and azimuth from plot center. Total height and height to the crown base were measured to the nearest decimeter on every fifth tree. Plot descriptors were evaluated within a circle of 300 m². Elevation is measured to the nearest 100 m, slope is measured to the nearest 10%, and aspect to the nearest 45°. Additional site descriptors were measured but not used in this comparison study.

Plots were remeasured from 1986 to 1990, 5 years after establishment. In the remeasurement, diameter was recorded for every tree within a given plot, and height only on every fifth tree. No height to the crown base was recorded. Because the development of a height increment model requires repeated tree heights, 7,797 Norway spruce trees from 4,135 different permanent sample plots are available throughout Austria. Please note that all stand characteristics such as CCF and BAL are based on the total number of trees for a given plot, while the tree characteristics are only from the 7,797 trees with repeated tree heights.

RESULTS

Coefficient Estimates by OLS vs. 3SLS

Using the SYSLIN procedure in the Econometrics-Time Series module of SAS (SAS Institute, 1988), the parameters in Equations 1 to 3 were first estimated independently by applying ordinary least squares (OLS) and then simultaneously by using 2SLS and 3SLS.

Attention was immediately focused on the $ln(C)$ and CCF terms in the height increment model (2). These two terms were both significant (α=0.05) with OLS. With 2SLS, the $ln(C)$ term remained significant but the CCF term became strongly nonsignificant. With 3SLS, both the $ln(C)$ and CCF terms were nonsignificant. Thus the height increment model (2) was reduced from 8 parameters to 6.

$$\ln(\Delta H) = a + b_1 \cdot \ln(D) + b_2 \cdot H^2 + c_1 \cdot BAL + d_1 \cdot ELEV^2 + d_2 \cdot SL + \epsilon_2 \qquad (14)$$

The size of the matrix partitions in the variance-covariance matrix in (11) is reduced accordingly. Now $\tilde{C}_{22}$ is a 6×6 variance-covariance matrix for $\tilde{Y}_2$, and the off-diagonal partitions $\tilde{C}'_{12}$ and $\tilde{C}'_{23}$ are 8×6 and 6×10 matrices containing the cross-equation covariances between the estimated parameters, respectively. Because there are now $p_1+p_2+p_3$= 8+6+10=24 estimated parameters in the three growth equations, there are 576 variance-covariance elements in (11), a reduction of 100 elements.

Correlation between Predictions

The customary OLS approach with single tree modeling assumes independent increment equations. If this assumption is correct, then all elements of the covariance partitions $\tilde{C}_{12}$, $\tilde{C}_{13}$, and $\tilde{C}_{23}$ in Equation 11 equal zero. Because these covariance estimates among coefficients may be positive or negative, the combined effect for a given observation is not apparent until the resulting cross-correlation between predictions is calculated.

To examine the interrelationships among the equations in our system, we calculate the correlations between the predictions of *ln(BAI)*, *ln(ΔH)*, and *ln(C)* (models 1, 14, and 3) for each observation according to Equation 13. Figure 14.1 displays the cross-equation correlations $\tilde{r}_{ij}$ between each pair of predictions from models *i* and *j* vs. the respective predicted variables $\tilde{Y}_i$. In Figure 14.1, $\tilde{r}_{12}$ is massed around 0.23, with a maximum correlation of 0.33 between the predictions for *ln(BAI)* ($\tilde{y}_1$) and *ln(ΔH)* ($\tilde{y}_2$). This indicates that the first two equations in the system are fairly interdependent. The correlations $\tilde{r}_{13}$ between predictions from the first ($\tilde{y}_1$) and third equations ($\tilde{y}_3$) are weak, with a mean of –0.07, and extrema at 0.05 and –0.11. Although the correlations $\tilde{r}_{23}$ between predictions from the second ($\tilde{y}_2$) and third equations ($\tilde{y}_3$) have a mean of zero (0.01), they are more greatly dispersed than those for $\tilde{r}_{12}$, ranging from 0.26 to –0.10 with a standard deviation of 0.08.

In Figure 14.2 the same cross-equation correlations are first averaged by size class and then displayed against the major predictor variables: diameter at breast height (*D*), crown ratio (*CR*), crown competition factor (*CCF*), and basal area of the larger trees (*BAL*). The most interesting correlations are $\tilde{r}_{12}$ between the predictions of the first two growth equations. The mean correlation of $\tilde{r}_{12}$ vs. *D* is not constant, and exhibits a downward trend (Figure 14.2). For diameters < 50 cm, the correlation $\tilde{r}_{12}$ is constant at 0.23, but for larger diameter trees the correlation diminishes to 0.10 for 100-cm diameter trees. Similar behavior occurs with respect to *CCF*: $\tilde{r}_{12}$ is constant at densities below *CCF*=300, and declines with increasing density. For both *CR* and *BAL* the trend is positive rather than negative.

Figure 14.2 also illustrates the mean correlations for $\tilde{r}_{13}$ and $\tilde{r}_{23}$ vs. the same predictor variables. Essentially, there is no real trend, except that $\tilde{r}_{23}$ does increase regularly with *CR* and decrease with increasing *BAL*.

DISCUSSION

Because individual tree forest growth models are based on multivariate attributes observed on the same individuals, the resulting set of growth equations can be considered as a simultaneous system. Therefore, joint regression techniques should be considered for simultaneously estimating parameters. If endogenous variables do not appear on the RHS, the seemingly unrelated regression procedure of Zellner (1962) will improve the efficiency of the parameter estimates. If endogenous variables are used as predictor variables then multistage estimation techniques (2SLS or 3SLS) are necessary to obtain parameter estimates that are consistent; such estimates will also be asymptotically efficient. Ignoring simultaneous interactions by separately applying OLS to each model in the system can result in estimates that are biased and inconsistent.

Crown ratio (CR), a size variable and not an increment, is observed for the first period. Thus, we do not need to predict it to start with our three-equations. Our system is recursive and for the first step of a simulation, we could instead use the OLS predictor of the basal area increment model, rather than the 3SLS predictor. However, with increasing number of simulation runs the simultaneous estimation of growth equations becomes more and more important because it allows us to mimic nature more closely.

Using 3SLS with our system of equations allowed for deleting two nonsignificant terms that had not been detected under OLS. Deleting these two terms reduced the size of the

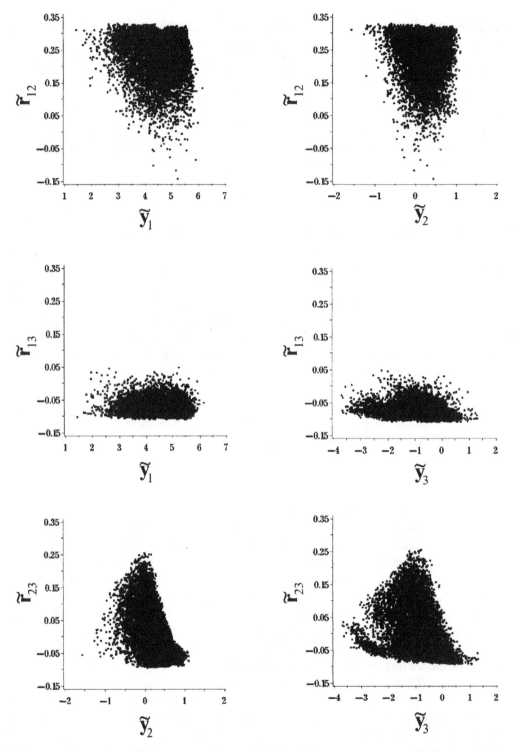

Figure 14.1. The cross-equation correlations $\tilde{r}_{ij}$ between each pair of predictions versus the 3SLS predictions. $\tilde{y}_1$ indicates the logarithm of the basal area increment predictions, $\tilde{y}_2$ the logarithm of the height increment predictions and $\tilde{y}_3$ the logarithmic transformation of the logistic crown ratio model $((CR/1)-1)$.

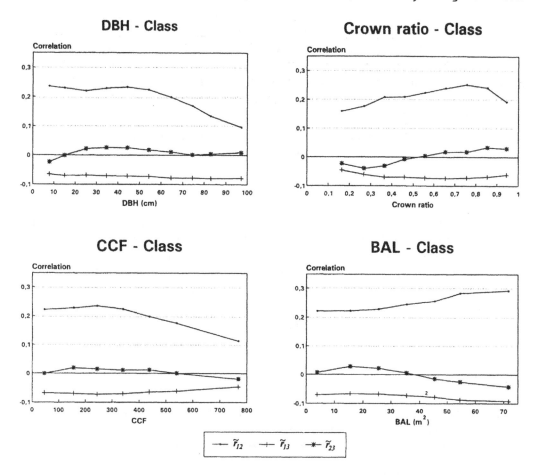

Figure 14.2. Relationships between the cross-equation correlations ($\tilde{r}_{12}$, $\tilde{r}_{13}$, and $\tilde{r}_{23}$) and the major predictor variables (by size class): diameter at breast height D, crown ratio CR, crown competition factor CCF, and basal area in larger diameter trees BAL.

variance-covariance matrix by 100 elements. Because one of the deleted variables was endogenous on the RHS, the simultaneous structure of the system was simplified. Consequently, the height increment model is no longer directly influenced by the crown model. Recognizing that the system is simultaneous allows for a more precise determination of the form of the component models.

Even though it is clear that a certain system is simultaneous, there are situations when simultaneous system estimation methods cannot be used. Based on theoretical considerations, Stage (1975) developed a height increment model that used diameter increment as a predictor variable. When coupled with the basal area increment model in the Stand Prognosis Model (Wykoff et al., 1982), the closely related diameter increment is nearly an endogenous variable on the RHS of the height increment model. It is not possible to estimate this system of equations simultaneously because the observations of the height increment came from a felled-tree data set that was independent of the basal area increment data.

One of the advantages of simultaneous regression techniques is that the asymptotic covariances of coefficients among equations can be estimated. Figures 14.1 and 14.2 indicate that cross-equation correlation between the predictions depends on the degree of dependency in the system. Predictions between the basal area and height increment model and between the height increment and the crown model are fairly well correlated, while correlations between predictions from the basal area increment and crown model are rather weak. Furthermore, cross-equation correlations such as $\tilde{r}_{12}$ are not constant and can vary with the value of the independent variables. As a result, the efficiency of 3SLS vs. OLS can be strongly influenced by the values of the predictor variables, as reflected in the estimated variance-covariance matrix for the full system of equations.

ACKNOWLEDGMENTS

This research was conducted when Hubert Hasenauer was a Visiting Scientist at the College of Forestry and Wildlife Resources, Virginia Polytechnic Institute and State University in Blacksburg, Virginia, and at the Intermountain Research Station's Forestry Sciences Laboratory in Moscow, Idaho. Hasenauer was working on a Schrödinger research grant from the Austrian Science Foundation. We are grateful to Karl Schieler and Klemens Schadauer of the Federal Forest Research Center in Vienna for making the Forest Inventory data available. Helpful review comments were provided by Hubert Sterba and Albert R. Stage and the anonymous referees.

REFERENCES

Aitken, A.C. On the least-squares and linear combination of observations. *Proceedings of the Royal Society of Edinburgh.* 55, pp. 42–48, 1934–1935.

Bitterlich, W. Die Winkelzählprobe. *Allgemeine Forst- und Holzwirtschaftliche Zeitung.* 59(1/2), pp. 4–5, 1948.

Borders, B.E. and R.L. Bailey. A compatible system of growth and yield equations for slash pine fitted with restricted three-stage least squares. *For. Sci.* 32, pp. 185–201, 1986.

Burkhart, H.E., K.D. Farrar, R.A. Amateis, and R.F. Daniels. Simulation of individual tree growth and stand development in loblolly pine plantations on cutover, site-prepared areas, Coll. For. and Wildlife Resources, Virg. Tech. Inst., Blacksburg, VA, Publication FWS-1-87, 1987.

Dixon, R.K., R.S. Meldahl, G.A. Ruark, and W.G. Warren. *Process Modeling of Forest Growth Responses to Environmental Stress.* Timber Press, Portland, OR, 1990.

Forstliche Bundesversuchsanstalt. Instruktionen für die Feldarbeit der Österreichischen Forstinventur 1981–1985. Forstliche Bundesversuchsanstalt, Wien, 1981.

Goldberger, A.S. *Econometric Theory,* John Wiley & Sons, New York, 1964.

Gregoire, T.G. Generalized error structure for forestry yield models, *For. Sci.* 33, pp. 423–444, 1987.

Haavelmo, T. The statistical implications of a system of simultaneous equations, *Econometrica* 11, pp. 1–12, 1943.

Hasenauer, H. Ein Einzelbaumwachstumssimulator für Ungleichaltrige Fichten-Kiefern-und Buchen-Fichtenmischbestände, Forstl. Schriftenreihe, Univ. f. Bodenkultur, Wien. Österr. Ges. f. Waldökosystemforschung und Experimentelle Baumforschung an der Univ. f. Bodenkultur. Band 8, 1994.

Hasenauer, H. and R.A. Monserud. A crown ratio model for Austrian forests, *For. Ecol. Manage.* 84, pp. 49–60, 1996.

Hasenauer, H. and R.A. Monserud. Biased statistics from height increment predictions. *Ecol. Modeling* 98, pp. 13–22, 1997.

Hasenauer, H., R.A. Manserud, and T.G. Gregoire. Using simultaneous regression techniques with individual tree growth models, *For. Sci.* 44, pp. 87–95, 1998.

Johnston, J. *Econometric Methods,* 3rd ed., McGraw-Hill Book Co., New York, 1984.

Judge, G.G., W.E. Griffiths, R.C. Hill, and T.C. Lee. *The Theory and Practice of Econometrics,* John Wiley & Sons, Inc., New York, 1980.

Kmenta, J. *Elements of Econometrics,* 2nd ed., The MacMillan Co., New York, 1986.

Krajicek, J.E., K.A. Brinkman, and S.F. Gingrich. Crown competition: A measure of density, *For. Sci.* 7, pp. 35–42, 1961.

Monserud, R.A. Methodology for Simulating Wisconsin Northern Hardwood Stand Dynamics, Thesis presented to the University of Wisconsin, Madison, in partial fulfillment of requirements for the degree of Doctor of Philosophy, 1975.

Monserud, R.A. and H. Sterba. A basal area increment model for individual trees growing in even- and uneven-aged forest stands in Austria, *For. Ecol. Manage.* 80, pp. 57–80, 1996.

SAS-Institute. SAS/ETS User's Guide, Version 6, Cary, NC, 1988.

Stage, A.R. Prediction of Height Increment for Models of Forest Growth, USDA For. Serv. Res. Paper INT-164, 1975.

Theil, H. Repeated Least-Squares Applied to Complete Equation Systems, The Hague: The Central Planning Bureau, The Netherlands, 1953.

Wykoff, W.R., N.L. Crookston, and A.R. Stage. User's Guide to the Stand Prognosis Model, USDA For. Serv. GTR INT-133, 1982.

Zellner, A. An Efficient method of estimating seemingly unrelated regressions and tests for aggregation bias, *J. Am. Stat. Assoc.* 57, pp. 348–368, 1962.

Zellner, A. and H. Theil. Three-stage least squares: simultaneous estimation of simultaneous equations, *Econometrica* 30, pp. 54–78, 1962.

Chapter 15

Incorporating Soil Variability into a Spatially Distributed Model of Percolate Accounting

Andrew S. Rogowski

INTRODUCTION

To describe behavior of natural systems in a Geographic Information Systems (GIS) framework we often rely on field observations and output from numerical models. Field observations are generally made at a point, while numerical models attempt to aggregate local point responses over an area in time. When modeling soil water status and distribution of recharge at a farm, watershed, or regional scale, point observations considered representative of an area, or volume of soil, usually serve as input to a numerical process model. When field observations are not available, modelers often use information from other sources as input to a process model. The principal source of information is soil surveys in a form of published maps and ancillary tabular data of soil properties. Variables derived from such surveys are assumed to apply uniformly to an entire area mapped as a given soil type and expressed on a map as a soil polygon accompanied by descriptive and laboratory data in a text form. In truth however, most variables associated with a given soil type represent observations made at a site, which although typical (*modal*) for a given mapping unit, may be far removed from the local study area. Moreover soil properties vary spatially and in time, and soil polygons themselves are far from pure and may contain *inclusions* of associated materials (Rogowski and Wolf, 1994). In addition, as Stein et al. (1991) point out, when operating in the domain of GIS and process modeling, there is always a choice of either: *(1) interpolating the input parameters first, and modeling the process later*, or *(2) modeling the process at a point first, and interpolating the outputs later.*

Different flow-related parameters vary differently in space and some are more stable in time than others. For example, values of soil bulk density tend to vary less in space and be much more stable in time than hydraulic conductivity. Moreover, flow models are often configured in one dimension to specifically accommodate one component such as runoff, infiltration, interflow, or groundwater recharge. My results suggest that when multiple parameters are needed as an input to a model, modeling of the process at a point where all of the input parameters are known or measured at the same time may be preferable to interpolating the individual parameters over the area first and modeling the process later (Rogowski, 1996b). The correctness of prediction, at least for the more dynamic outputs, appears to depend more on a number of nearby data points and model sensitivity to specific input parameters, than on the order in which interpolation and modeling are implemented (Rogowski and Hoover, 1996). Stein et al. (1991) likewise found that simulation of water flux at the points where water retention curves *(WRC)* have been measured followed by interpolation between locations, gave a more realistic distributions of flux, than interpolation of *WRC* followed by simulation. However, Sinowski et al. (1997) and De Gruijter et al. (1997) argue for interpolation first followed by simulation. We may note that both of the latter studies use few simple and relatively static parameters characteristic of a mapping unit. Under these circumstances, interpolation first followed by simulation may well be superior. However, when modeling dynamic processes taking place in a soil matrix such as water flux, or re-charge distribution, which require a large number of dynamic variables, this approach will lead to overparametrization. Since model sensitivity may also be time and scale dependent, the choice of approach depends on the complexity and purpose of the model. When mul-tiple, scale-dependent parameters, which also vary in time, are used as input to a dynamic deterministic or stochastic model, simulation first followed by interpolation makes much better sense, especially when a model itself is configured in one dimension. There are, of course, exceptions. Quite often, as I demonstrate later in this chapter, some parameters (e.g., elevation, slope, aspect) acquired through remote sensing are available for the entire area, their resolution limited only by the resolution of the sensor. In that case, interpolation of additional input variables first followed by process modeling is to be preferred.

To predict recharge I have used a deterministic flow model defined as a one-dimensional water budget in a soil column. The model computes daily estimates of surface runoff, layer by layer soil water content, actual evapotranspiration *(ET)*, plant water stress, and deep percolation (recharge flux), based on environmental, biological, and mechanical state of the control volume (i.e., soil column). It calls for input of weather data and multiple soil and plant variables, some of which may change daily. In that setting, a preferred use is the second option of modeling the soil water distribution and the vertical flux components at a point for each location first, and then distributing them spatially over all GIS nodes of the study area. The alternative is less operationally attractive, since it would require modeling and estimation of multiple crop and soil model parameters for all nodes, followed by a time-intensive simulation of water balance at each node. Because observed soil water prop-erties vary in time and spatially as a function of scale, landform position, and soil type, actual values at a new location can never be predicted exactly, and each prediction carries with it a measure of associated uncertainty (Oreskes et al., 1994).

My objective here is to demonstrate a method of incorporating uncertainty associated with more dynamic soil attributes into a model of percolate accounting, and to illustrate how the model performs at a catchment scale. The methods and *tools* I use are well known in literature. The *novelty* of the approach lies in how different tools, in this case a determin-istic flow model, a stochastic (geostatistical) simulation, and a GIS framework, are put together and applied to solve a specific problem of modeling the distribution of soil water content, recharge, and interflow at a scale of the Case Study site (1.2×1.2 km) in time.

THE MODEL

The Soil-Plant-Atmosphere-Water (*SPAW*) simulation model (Saxton et al., 1992; Saxton, 1994) estimates evapotranspiration *(ET)*, plant water stress, and layer-by-layer changes in soil water content. The outputs are recharge, modeled as the percolate flux from below the root zone (q_v), and the distribution of soil water content by volume θ_v. The model is designed to provide a daily, one-dimensional (vertical) water budget at a point in a soil column. Actual measurements of *ET,* soil water content θ_v, and other input variables may be substituted for default values, or periodically used to verify or recalibrate the model. The model requires daily climatic and plant growth estimates, and a detailed in situ horizon-by-horizon description of soil profile, including either pertinent soil moisture characteristics, or textural classification information for each layer.

Interaction between daily precipitation *(P)* and evapotranspiration *(ET)* drives the model. In the Case Study described here, daily precipitation was measured at the site in a recording rain gauge, and potential evapotranspiration *(PET)* was estimated from pan evaporation at the Met Site located down the valley, 7 km away. The *PET* values were used to compute actual *ET,* which was then subtracted from available soil water in storage. The model takes observed values of *P* corrects them for vegetation interception and runoff, and distributes the infiltrate among soil layers to storage by *cascading* it down from the surface. Available water in each layer is subject to daily *ET* withdrawal depending on a crop, stage of growth, and root distribution. The amount of water which reaches the bottom boundary layer becomes the vertical percolation recharge flux (q_v). Figure 15.1 illustrates a typical partitioning of infiltrate between *ET* and q_v for one location from March 15th to November 30th, 1990. Adjustments among layers due to unsaturated water flow by redistribution are made using a finite element subroutine based on Darcy's Law, soil moisture characteristics, and hydraulic conductivity values computed from texture. The output consists of daily summaries of soil water θ_v for each layer, in addition to a daily digital and graphical output of precipitation *P, ET,* and percolation flux q_v. The model addresses the soil-plant-air continuum at a point, and output estimates are made conditional to the specific profile hydrologic properties, as well as the local climatic and biotic factors.

The *SPAW* model uses soil texture class, sand, and clay content values to compute the necessary hydrologic input parameters. The texture classes are generally part of soil descriptions for soil types identified on a soil map. Once a texture class is known, sand and clay content values can be obtained from a soil database. In this case I have used the Pennsylvania State University (PSU) Soils Database (Ciolkosz and Thurman, 1992). The PSU Database is a compilation of information on 800 pedons representing 170 different soil mapping units in Pennsylvania that have been collected and analyzed over the years by the PSU soil characterization laboratory. There are generally several descriptions of the same mapping unit sampled at different locations in the state. The database contains detailed profile and site descriptions, as well as layer-by-layer physical, chemical, and mineralogical information. Distributions of paired values of pressure head ψ and water content θ_v, and θ_v and hydraulic conductivity K_v were estimated for this study for each soil layer from texture (Saxton et al., 1986), and expressed as soil moisture characteristic $(\psi [\theta_v])$, hydraulic conductivity $(K_v[\theta_v])$, and soil water content at field capacity (θ_{fc}). Hydraulic conductivity at field saturation was taken as the last pair of values from the moisture characteristic and hydraulic conductivity curves as $\psi \to 0$, while θ_{fc} was computed from Saxton (1994) as,

$$\theta_{fc} = 0.30 - 0.0023 \text{Sand}\% + 0.005 \text{Clay}\% \qquad (1)$$

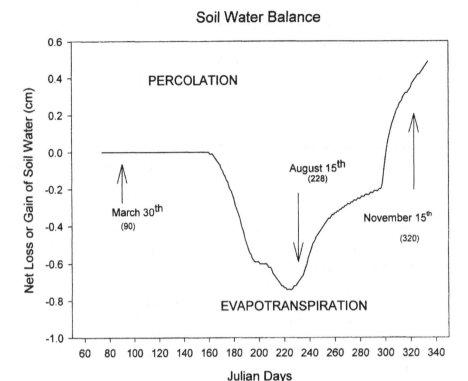

Figure 15.1. Partitioning of incoming precipitation between recharge modeled as a vertical percolation flux q_v (positive) and evapotranspiration *ET* (negative), from March 15th to November 30th, 1990 for a single location; arrows indicate dates illustrated in this study.

CASE STUDY

The Case Study site was on the Mahantango Creek watershed in Pennsylvania, USA, (Rogowski and Wolf, 1994). The watershed is situated in the Valley and Ridge physiographic province of the Appalachian Mountains, characterized by varying relief: upland hills, valleys, and forested mountain ridges dissected by streams. The climate is humid with rainfall distributed evenly, at the scale of the study area (Rogowski, 1996b), both spatially and in time (about 1000 mm/yr). Elevation ranges from less than 100 m in the valleys to over 500 m on the ridge tops. The ridges, valleys, and streams are oriented northeast-southwest, corresponding to the regional strike of major rock formations, while beds dip to the northwest and southeast along the centrally located anticline.

The land use is predominately cropland in a rotation management system of corn, small grains, and meadow, intermingled with numerous tracts of woodland, areas of permanent pasture, and orchards. In here, estimates are as if the whole area were in corn or pasture, during a spring to fall growing season. These two alternate land uses were selected for demonstration because they tend to affect recharge and water use differently. The model assumes corn produces roots down only to the 0.5 m depth, while on pasture root distribution is allowed to vary from 0.5 to 1.0 m depth depending on soil type and stage of growth. The Case Study area is 1.2 km on the side with individual cell size of 5×5 m. The site is part of an area where a detailed DEM derived from aerial stereo-photography was commercially acquired. The DEM has a vertical resolution of 0.1 m with maximum vertical error estimated as <0.5 m (except under closed canopy) and horizontal resolution of 5 m, densified from 15 m (White, 1997). The DEM, used to compute the Slope and flow direction vector (Aspect) for each 5×5 m cell, was overlaid with a detailed soil map in the GIS *IDRISI* (Eastman, 1992). Plate 9 shows the corresponding maps of Elevation (a), Aspect (b), Slope (c), and Soil Type (d) at the Case Study site. The points on a Soil Type map indicate locations of the soil pedons used as input to the model.

For the eight soils types found at the Case Study site (Plate 9d), the PSU Soils Database included about 60 individual pedon descriptions. Since most mapping units contain small areas of associated soils, the 60 individual pedons were augmented with about 40 pedons of associated series, considered as likely inclusions. Mimicking a field procedure, pedon locations at the study site were picked at random by an *IDRISI* module to cover the study area as evenly as possible, and avoid clustering (Figure 15.2). Their coordinates were then read off, and a soil polygon mask (Plate 9d) was intersected with the overlay of locations. Tabulated pedons of a given mapping unit from the PSU Soils Database were assigned to the locations within a polygon of the same soil type until no more remained. Additional assignments were made from pedons representing associated inclusions. Figure 15.2 shows the original locations of pedons in the state and the subsequent assigned locations within the Case Study area, while Plate 9d indicates the position of assigned pedons relative to the soil type mapping units. To verify that the structural similarity of the respective distributions between the *State* and *Site* locations was preserved, I performed the structural analysis on both the initial *State* and final *Site* locations in terms of *SPAW* generated q_v values. Figure 15.3 shows the respective distributions of q_v on March 30th, August 15th, and November 15th for the 101 pedons, and Figure 15.4 illustrates the raw (dashed) and modeled (solid) semivariograms (Table 15.1), for the August 15th distribution. The histograms represent distributions and parameter statistics of q_v for different dates, while the semivariograms illustrate the similarity of spatial structure on a single date. Both semivariograms have essentially the same sill and expected differences in the range and nugget effect. Differences in the nugget suggest more unexplained variability at the larger *State* scale, while the *State* range is approximately equal to the *Site* range squared.

FLOW MODELING

The *SPAW* model was run on each pedon profile from March 15th through November 30th. The run was started with a uniform water content (θ_v) of 0.20 cm³/cm³, mimicking uniform spring conditions, which subsequently were adjusted to volumetric field capacity θ_{fc} for each layer (Equation 15.1). Field-saturated hydraulic conductivity of each layer also was estimated from the 10 textural classes specified in Saxton (1994). To define site anisotropy (Zaslavsky and Rogowski, 1969), vertical and horizontal field-saturated hydraulic conductivity for the soil profile as a whole had to be computed. The vertical field-saturated hydraulic conductivity (K_v) was computed as a harmonic mean of component layers weighted by a layer thickness. The horizontal field-saturated hydraulic conductivity (K_h) was assumed to equal the conductivity of the top layer if the layer below it had a lower K_{fs}. If not, K_h was taken as an arithmetic average of component layers weighted by the layer thickness. The ratio, $n = K_h/K_v$, was assumed to represent profile anisotropy. Given a value of vertical

Pennsylvania, USA

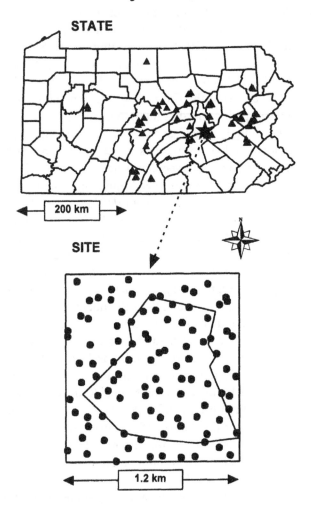

Figure 15.2. Location of the 60 soil pedons on the map of Pennsylvania and the assigned locations of 101 pedons (including 41 inclusions) at the Case Study site showing the catchment boundary; the star on the state map indicates location of the study site.

percolation flux q_v, the interflow vector at a point can be approximated as a horizontal flow component q_h for a specified cell direction β (Aspect from Plate 9b),

$$q_h(\beta) = q_v \times n \times tan\ \alpha \qquad (2)$$

where α represents land slope in % from the 5 m DEM (Plate 9a), and $tan\ \alpha$ is equivalent to the (slope gradient)$^{-1}$ definition in *IDRISI*.

The *SPAW* model simulates a continuous daily water budget accounting. Tabulated output lists the daily amounts of precipitation, runoff, recharge flux q_v and evapotranspiration (*ET*), along with volumetric soil moisture content θ_v for each soil layer. Under our experimental conditions there was very little runoff. A previously cited example in Figure 15.1

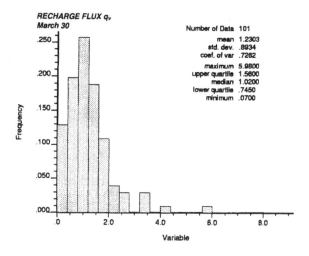

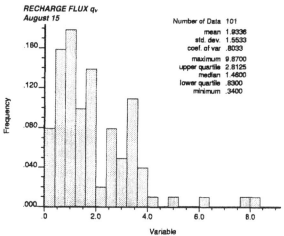

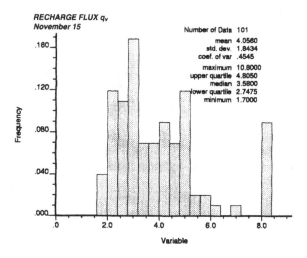

Figure 15.3. Histograms of recharge flux q_v distribution for 101 locations on March 15th, August 15th, and November 15th computed from the Soil-Plant-Atmosphere-Water (*SPAW*) model (Saxton, 1994).

Semivariance of q vertical

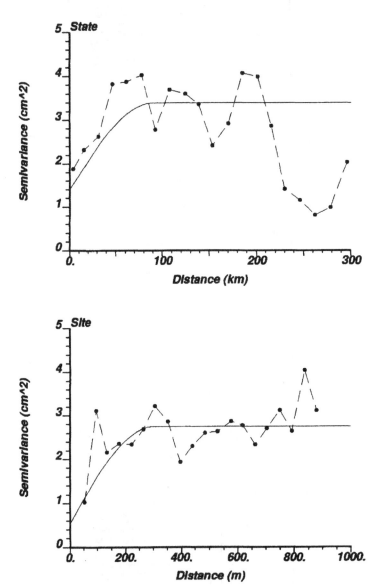

Figure 15.4. Raw semivariograms (dashed) of the computed recharge flux q_v for the statewide distribution of 60 pedons (*State*), and for the Case Study site assigned distribution of 101 pedons (*Site*); the corresponding spherical models are shown as a solid lines, and their parameters are listed in Table 15.1.

illustrates for a single location a distribution of mass balance in time, as a function of *ET* (negative) and q_v (positive).

A rectangular correlation matrix of Pearson correlation coefficients was generated using *SPAW* output for all pedons (101) used as input (SAS, 1995). Results suggested that moisture contents θ_v in adjoining layers should be reasonably well correlated with each other

Table 15.1. Comparison of Structural (Semi-variogram) Parameters of Computed[a] Deep Percolation Flux q_v for the Actual *State* and Assigned *Site* Distribution of Pedon Locations on August 15th

Parameter	State	Site
Nugget C_0 (cm²)	1.40	0.55
Sill C_1 (cm²)	2.00	2.20
Range ρ (m)	$(300)^2$	300

[a] Computed from SPAW model (Saxton, 1994).

(ρ=0.71 to 0.85), but showed little or no correlation among layers that were further apart. This was not unexpected, since the flow subroutine in *SPAW* cascades soil water from layer to layer based on the head difference between adjoining layers and the mean value of their conductivity. The results also showed that moisture contents θ_v in a top layer were negatively correlated with *ET* (–0.74), while those in the 4th and 6th soil layers tended to be positively correlated (0.72) with the recharge flux q_v. Predictably, soil water evaporation in the model is generally removed from the first evaporative layer, while the 4th and 6th layers, respectively, reflect a lower boundary of shallower (Calvin, Berks) and deeper (Meckesville, Leck Kill, Watson) soil types.

In the applications mode, each day's output from *SPAW* for every location can be realized separately as a spatial overlay in GIS by applying either kriging to the values at assigned locations, or by using these values as conditioning data in a stochastic simulation procedure. Kriging leads to a single smoothed-out interpolated realization, while the stochastic imagining results in a series of alternate maps consistent with the conditioning data and the correlation model (Journel, 1996). When both the kriging and simulation are performed on the same data, a kriged distribution is in general quite similar to a smoothed-out average of a large number of stochastic realizations (Journel, 1989, 1996). This should also hold for highly correlated variables if measured at the same locations. As the correlation decreases, the correspondence will decline. Because the 101 values of θ_v in the 4th and 6th soil layers tended to be only weakly (0.72) correlated with the recharge flux q_v, only a weak relationship between kriged overlays of θ_v and simulated overlays of q_v may be expected. To obtain daily overlays of soil water content θ_v which varied little, I have used kriging *(OK)*.[1] For parameters that were quite variable and likely to contain significant outliers, such as percolate flux q_v and anisotropy ratio n of the field saturated hydraulic conductivities K_h/K_v, I have used a sequential conditional simulation procedure *sisim*;[2] (Rogowski, 1996a) to generate a number (25) of equiprobable (Journel, 1995) realizations.

Finally, to model spatial distribution of interflow as a horizontal flux component q_h in the direction β, I have used a mixed approach alluded to earlier. I first modeled q_v and n using *SPAW* generated values of q_v and prior knowledge of profile anisotropy (K_h/K_v),

---

[1] Computed using ordinary kriging *(OK)* program from Deutsch and Journel (1992, 1997).

[2] Sequential indicator program *sisim* (Deutsch and Journel, 1992, 1997).

followed by conditional simulation of their spatial distributions, which were subsequently combined (Equation 2) in GIS with spatial distributions of flow direction β (Aspect—Plate 9b) and *tan* α to give an overlay of interflow as a spatial distribution of $q_h (\beta)$. The four overlays that were constructed in the GIS *IDRISI* and then had to be combined according to Equation 2 were:

- conditionally simulated distribution of recharge as vertical percolate flux q_v
- conditionally simulated profile anisotropy $n = K_H/K_v$
- distribution of the flow direction β—Plate 9b
- distribution of slope gradient *tan* α—from Plate 9c.

CONDITIONAL SIMULATION

Srivastava (1996) presents a general overview of different conditional simulation procedures including multi-Gaussian and the indicator sequential simulation approaches. Although both types of methods utilize a similar algorithm, they differ in how the local conditional probability distribution (lcpd) is estimated. The usual practice is to choose a node to be simulated, estimate the lcpd, draw a random value from that lcpd, include it in the set of conditioning data, and repeat the steps until all nodes have been filled. However, to estimate the lcpd, multi-Gaussian methods assume a normal distribution with a known mean and standard deviation, while indicator kriging makes no such assumption regarding the shape of the distribution, and develops it by construction; i.e., by estimating the probability that a value at a given node is below a specified cutoff for a number of thresholds.

Results in literature (Journel and Alabert, 1989) suggest that a conditional simulation approach known as the sequential indicator simulation (*sisim*) will reproduce spatial patterns of continuity and can also be used to model spatial uncertainty (Gomez-Hernandez and Srivastava, 1990; Deutsch and Journel, 1992, 1997; Olea, 1991, p. 71; Goovaerts, 1997). The principal advantage of *sisim* over the Gaussian-related algorithms is *sisim's* ability to control a number of associated spatial covariances instead of assuming a single covariance model for the whole distribution (Journel, 1989, p. 35). Perhaps even more critical is the maximum entropy property of the multi-Gaussian random function models that does not allow for correlation of extreme values (Journel and Deutsch, 1993), and tends to maximize their scattering in space (Goovaerts, 1997, p. 393). Although Cressie (1991, p. 281) suggests that indicator kriging (which is part of *sisim*) theoretically gives a worse approximation of the conditional expectation than disjunctive kriging, the latter may be viewed as a generalization of a bivariate Gaussian model with the associated normality assumptions. However, the normality aspect of multi-Gaussian models should be a property of the data and not of the model. Thus prior rescaling of the data with a normal score transform does not necessarily solve the inherent problem, if the data themselves are not normally distributed (Deutsch and Journel, 1997, pp. 87, 143). In contrast, the *sisim* approach is nonparametric. Since *sisim* discretizes a distribution into a number of thresholds each modeled with a discrete semivariogram, it enables one to test for continuity, or *connectivity* specific to different classes of values found in the distribution (Goovaerts, 1997). My purpose in using *sisim* was to construct from the conditioning data multiple equiprobable spatial "realizations," or outcomes, of attributes q_v and n that would reproduce spatial patterns of variability and illustrate potential connectivity associated with the occurrence of extreme values of interflow modeled as horizontal flow q_h. Since input distributions—see, for example, the histograms of q_v, q_h, and n (Figure 15.5)—were nowhere near normal, the proper model to use was *sisim*.

To model different portions of a field distribution, the sequential indicator simulation program partitions the observed population of *SPAW*-derived q_v and n values into several classes, each separated by a threshold corresponding to the quartile of the distribution. For each observed value of q_v and n it then defines the indicator transformation with respect to that threshold. Thus for a random variable $Z(u)$ at a location u and a threshold value z, the corresponding indicator transformation $I(u;z)$ would be,

$$
\begin{aligned}
I(u;z) &= 1, & \text{if } Z(u) \leq z \\
&= 0, & \text{otherwise}
\end{aligned}
\tag{3}
$$

Given a set of q_v values, the *sisim* procedure converts them into indicator distributions for each threshold. It then defines an algorithm for adding an indicator value at the unsampled location which is consistent with a spatial structure (covariance) of the observed population for the particular threshold (Deutsch and Journel, 1992, 1997). The first step involves estimation of the conditional probability that a new value is less than a given threshold. This is done by kriging each unsampled location using the surrounding indicator values. The resulting estimate of the local conditional probability distribution (lcpd) is somewhere between 0 and 1. The actual simulation of a corresponding indicator value at the unsampled grid location is accomplished by drawing a random number between 0 and 1 from a uniform distribution. If this number is less than or equal to a kriged value, a simulated indicator value of 1 is assigned to that location. If it is more, a value of 0 is assigned. The new value becomes a part of the conditioning data set and the procedure is repeated at another location until all nodes in the catchment grid are filled.

To illustrate the principles involved, the soil water content θ_v, recharge flux q_v, and ET distributions derived from *SPAW* for 101 pedons were analyzed for three dates: March 30th, August 15th, and November 15th, 1990. These dates appear to be representative of the general mass balance distribution in time at this location (Figure 15.1). March 30th falls just after the start of the study, during a period of adjustment when profile storage fills up and no net ET or q_v occurs. August 15th is at the time of modeled peak ET demand, and November 15th refers to a period of decline in ET and increasing recharge.

To obtain a continuous coverage over the Case Study area, soil water content θ_v distributions were kriged (Figure 15.6, Plate 10), while those of recharge, modeled as vertical percolate flux q_v were conditionally simulated (Figure 15.7, Plate 11). Kriging θ_v gave smoothed-out distributions of soil moisture for the first, fourth, and sixth layer that show some variation in space and time, while conditional simulation better reflected the variability and uncertainty (Journel, 1996), in spatial and temporal patterns of recharge q_v and interflow q_h based on the simulated distributions of both the q_v and anisotropy n.

SOIL WATER

Plate 10 shows spatial distribution of soil water in the first (01), fourth (04), and sixth (06) layers on March 30th, August 15th, and November 15th based, on kriging the 101 q_v values for individual pedon layers derived from the *SPAW* simulation. Note that the scale

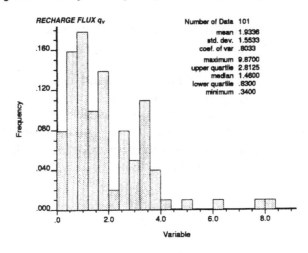

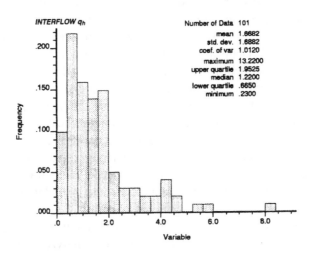

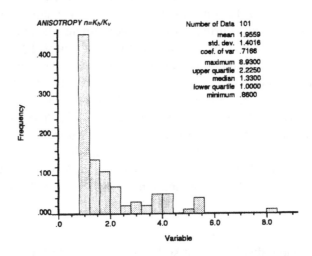

Figure 15.5. Histograms of the computed recharge flux q_v, interflow q_h, and the anisotropy K_h/K_v ratio n distributions on August 15th for the 101 locations shown in Plate 9d and Figure 15.2.

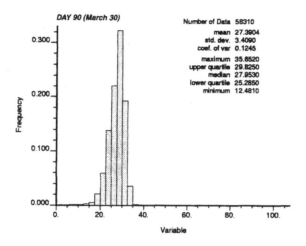

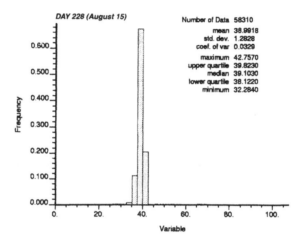

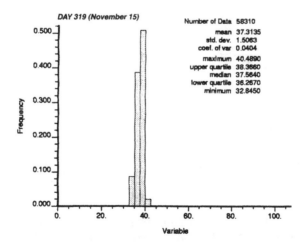

Figure 15.6. Histograms of the modeled *(OK¹)* distribution of the soil moisture content by volume θ_v in the surface layer (01) at the Case Study site on March 30th, August 15th, and November 15th.

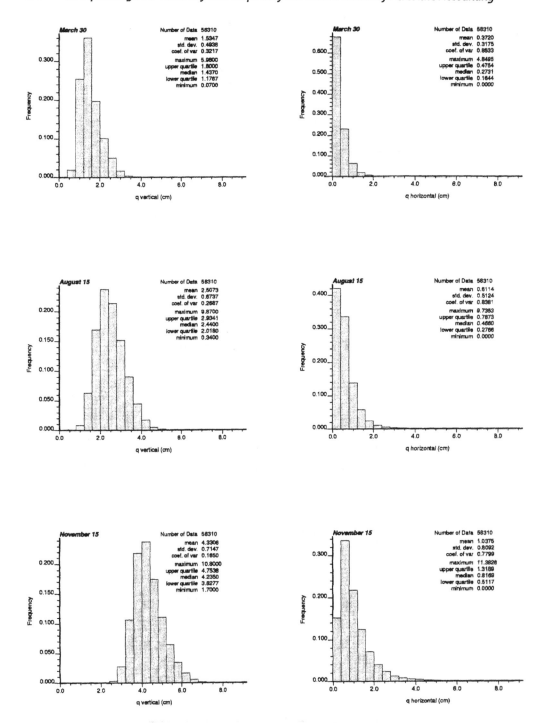

Figure 15.7. Histograms of the modeled distributions of recharge and interflow as vertical (q_v) and horizontal (q_h) percolate flux at the Case Study site on March 30th, August 15th, and November 15th.

for the March 30th date is different from that for the other two dates. This was a period of adjustment of the original θ_v to θ_{fc}. In the following period (August 15th) the geometry of the distribution was quite similar to that for March 30th but the space of uncertainty (i.e., the spread of the distribution in Figure 15.6) was considerably different, ranging from $\theta_v=0.32$ to 0.44 vs. $\theta_v=0.12$ to 0.36 by volume, for the March 30th date. The kriged maps of θ_v in the top layer generally reflect the distribution of *ET* (Saxton et al., 1992). Those for the 4th and 6th layer are expected to be weakly related to the average of stochastically simulated recharge flux q_v on the Case Study area (Journel, 1996).

Distributions of soil water in the first layer in Plate 10 are indicative of the spatial *ET* demand under corn. Since Pearson's correlation coefficient for the 101 *SPAW* generated values of θ_v in this case is negative (–0.74), low areas of kriged θ_v should correspond to zones with high cumulative evaporative demand and vice versa. Thus, deep blue areas on the Layer 01 map may indicate evaporative sinks on Mecksville soil (Plate 9d) which generally contain a flow-impeding fragipan layer, and thus may have less available profile water. Despite the lack of correlation among other than adjoining layers for the 101 *SPAW* generated values of θ_v, kriged distributions of soil water for the 4th and 6th layers in Plate 10 looked very similar to that for the 1st layer. That is, low θ_v areas tended to remain low throughout, and high θ_v areas tended to remain high. This suggests that water contents in all layers when subject to spatial *ET* demands at the surface and drainage flux q_v at the bottom of the root zone or impeding layer, may be correlated with depth throughout the profile. Indeed, as Ahuja et al. (1993) point out, the change in soil water content at the soil surface two days following a precipitation event may well be correlated with the hydraulic conductivity, steady state infiltration rate, and drainable porosity (van Schilfgaarde, 1957) of the soil profile.

Correlation analysis of q_v output from *SPAW* model for the 101 conditioning data pedons also suggests a possible correlation between recharge flux and soil water content in the shallower 4th layer (Calvin, Berks soils) and deeper 6th layer (Meckesville, Leck Kill, Watson soils). There is some visual evidence that the kriged distributions of θ_v in Plate 10 are related to the expected values (Plate 11), quartiles, and probabilities of exceedence (Plate 12) of the stochastically simulated flux q_v, but the relationship is neither clear-cut, nor conclusive. For example, low θ_v (deep blue) values in Plate 10 for the Layer 06 (November 15th) correspond roughly to lower q_v zones in Plate 11 for the same date. However, the upper deep blue zone in Plate 10 (low θ_v) appears to overlap part of a high q_v flow zone (red) in Plate 11 corresponding to the area where we might expect a horizontal q_h flow component in the SW direction. Higher θ_v values in Plate 10 tend to be located in areas with a higher q_v in Plate 11 and Plate 12 but not with the one-to-one correspondence. It appears that although soil moisture content and flux are related, when looking for an explanation at a specific location, distributions of other contributing factors, such as soil type, slope, and aspect, also need to be considered.

DISTRIBUTION OF FLOW

The output plots in Plate 11 show a distribution of *expected values* for the deep percolate flux q_v (cm) on each of the three dates (March 30th, August 15th, and November 15th). Distributions were computed with a *sisim* model for the catchment as a whole, and represent values based on 25 individual equiprobable realizations (Deutsch and Journel, 1992).

Equation 2 was then used to compute the horizontal flux component q_h from 25 realizations of $n = K_h/K_v$, expected values of q_v, and slope at each node (Plate 9). The computation of q_h illustrates application of the first approach of Stein et al. (1991), that of interpolating first, modeling second. Under the circumstances it was a simpler choice. In general, the mean of a finite number of simulations of q_v need not be equal to the *expected value;* however, for a declustered sampling of a distribution, arithmetic mean is generally used as an estimate of expected values (Olea, 1991, p. 24). Moreover, as the number of simulations increases, their mean usually converge to the expected value (Isaaks and Srivastava, 1989, p. 550). The illustrations here are based on 25 simulations; however, in practical applications 100 or more simulations should always be considered.

Simulated vertical flows (q_v) appeared to be consistently larger than their horizontal counterparts, except on steeper slopes. The flow distributions in Plate 11 initially show increased q_v just within the central zone. This more prominent recharge zone occurs at first in the middle of the catchment. As time goes on, the model suggests this zone may become more diffuse in the summer, and breaks up into separate prominent clusters in the fall when several primary recharge areas seem to be located in the south central, SW, and NW portions of the catchment on sections covered with younger and more permeable soil types (Calvin and Berks). The south central zone is also a zone of terrain depression and a natural flow pathway which connects that part of the catchment to the main watershed drainage.

Highest values of interflow q_h in Plate 11 (red) are associated, as expected, with the steepest slopes (25–60%), with significant contributions on slopes as low as 8–15%. Visually there seemed to be little difference in q_h among the three dates, because of the overriding influence of slope, thus Plate 11 shows q_h for November 15th only. However, a closer inspection of both the q_h and q_v distributions and their statistics in Figure 15.7 suggests that both the flux magnitude and the space of uncertainty associated with these distributions may increase in time. Thus areas with high q_v (~5 cm) may indicate zones of preferential recharge. Similarly, locations with above average q_h may indicate converging interflow, or seepage prone areas.

Both Plates 10 and 11 represent water content and flux distribution under corn. Results under pasture were very similar to those for corn, but spatial distributions of flux q_v on August 15th appeared more diffuse and less clustered than those in Plate 11, and recharge flux (q_v) on pasture exceeded that on corn by 20%, or more. Table 15.2 compares the statistics of conditioning data with those for the simulated values for the August 15th date on both land uses. The means of the expected values for both corn and pasture simulations appear to be consistently larger, though less variable, than those for the conditioning data.

RISK ASSESSMENT

One advantage of conditionally simulating a large number of realizations is the subsequent ability of the programs like *postsim* (Deutsch and Journel, 1992) to extract from such a set a number of summaries using the values to derive distribution statistics of a property over an area (Goovaerts, 1997, p. 431). To do so with a degree of confidence it is advisable to have a reasonably large number of realizations (simulations) to work with. Plate 11 shows the distributions of expected values of q_v for three dates. In contrast, Plate 12 illustrates spatial distribution for the three quartiles of q_v on a single date of August 15th and the probability (%) of a location specific q_v value exceeding, or being less than a given thresh-

Table 15.2. Comparison of Statistics for the Conditioning Data and Model Distributions for Vertical (q_v) and Horizontal (q_h) Recharge Flux Components under Corn or Pasture, and for the Degree of Anisotropy (n) on August 15th

Statistic	Data		SISIM		Data		SISIM		Data	SISIM
	q_v(cm)				q_h(cm)				$n=K_h/K_v$	
	Corn	Past.	Corn	Past.	Corn	Past.	Corn	Past.		
Number	101		58310		101		58310		101	58310
Mean	1.93	2.40	2.51	3.11	0.46	0.53	0.61	0.73	1.96	2.31
Std. Dev.	1.55	1.59	0.67	0.56	0.85	0.83	0.51	0.54	1.40	0.65
Max	9.87	10.7	9.87	10.7	7.11	7.73	9.74	10.6	8.93	8.93
Q3	2.81	3.29	2.93	3.47	0.49	0.54	0.79	0.93	2.23	2.71
Median	1.46	2.11	2.44	3.08	0.18	0.24	0.47	0.58	1.33	2.21
Q2	0.83	1.38	2.02	2.72	0.08	0.13	0.28	0.37	1.00	1.83
Min	0.34	0.64	0.34	0.64	0.02	0.09	0.00	0.00	0.86	0.86

old. To derive a quartile map, simulated values for each pixel were ranked and the quartiles *(Q)*, were selected as simulated pixel values corresponding to the 25th, 50th, and 75th percentile in the ranking. These pixels were subsequently combined into a single GIS overlay (Rogowski, 1996a). The three quartiles in Plate 12 are Q_1, median Q_2, and Q_3. The Q_1 realization focuses on highlighting the potential outliers, since the probability of a lower value than shown is only 25%. The median (Q_2) map tends to minimize the mean of absolute deviations for each pixel location; that is, individual pixel values are close to what they appear to be. In the Q_3 map the focus shifts to the reliable low zones, because in this case the probability of exceedance would be only 25%. This map and the Q_2 map are the closest in appearance to that in Plate 11. The results suggest that a high recharge zone is real and its occurrence persists in the quartile maps. Equally persistent are the low flow areas in the east and southeast corner and along the lower portion of western boundary of the study area.

Probability (*P*) maps (Plate 12) provide additional detail about potential existence and continuity patterns associated with either the low or high flow zones of q_v. The maps are expressed as percent probability that an area is either higher or lower than a given cutoff. The choice of a cutoff is arbitrary (Rogowski, 1996a). My maps here were prepared only as illustrations and were constructed for two cutoffs: $P \leq 5\%$ and $P \geq 95\%$. The $P \leq 5\%$ map delineates potential low flow zones, while the $P \geq 95\%$ map shows high flow areas in terms of the probability of occurrence. Both maps confirm the existence of a centrally located, connected, high q_v area and corresponding low recharge zones in the eastern and western portions of the catchment. In parts, the $P \leq 95\%$ map does reflect the kriged distributions of θ_v for the 4th and 6th layers. Areas with a larger probability of recharge flux greater than the 95% cutoff value appear to have a higher moisture content.

CONCLUDING REMARKS

Practical uses of the information presented in Plates 11 and 12 are numerous. We may utilize it, for example, in precision farming applications to monitor and predict nutrient losses. Delineating specific areas with a high probability of recharge flux may indicate sites vulnerable to groundwater pollution. Predicting soil moisture availability, and custom tailoring land use and management practices to potential recharge may be another area. For example, the *alley cropping* and *agroforestry* technology in developing countries (Kang and Wilson, 1987) would benefit from optimizing a selection of land areas suitable for a given plant species. The conditional simulation approach allows us to examine and draw conclusions from several sources simultaneously: from individual realizations, from expected values based on all the realizations, and from individual quartiles of simulated distributions. It also permits us to provide spatial estimates of a value at a specific location exceeding (or being less than) a chosen threshold. This feature is particularly attractive if the simulation results need to be explained to the decision-makers. In general, simulation models may become more realistic if their output were given in such probabilistic terms.

Variability of soil properties under natural field conditions is both time and space dependent. Potentially misleading interpretations may occur if variability of the spatially and temporarily distributed phenomena is interpreted only from regressions of correlated variables, or expressed in terms of means, and standard deviations of measured values for a single occurrence. This study attempts to remedy the situation by proposing a stochastic procedure for incorporating some aspects of soil variability into numerical models of water flow and nutrient transport distributed in time. When modeling in a GIS framework, users are faced with a choice of interpolating first, modeling later, or modeling first, interpolating later. The answer depends on the complexity of a model and availability of data. For example, as I have done here, to incorporate a measure of spatial variability into soil survey data, we can extract multiple pedon information from a database and assign it at random to locations within mapped polygons. If additional information is available from other sources, locations can be further stratified by topography, soil depth, drainage, and color. The key feature is that different kinds of complex information about a pedon can be utilized together, suitably weighted to emphasize most important aspects. For example, we can model daily water balance at each location for each pedon layer, extracting the deep percolate recharge flux q_v at a point. The obtained values become the conditioning data in a subsequent simulation to distribute q_v over a catchment, and the second choice of modeling first, interpolating later, is used.

Simulation results may suggest the existence and possible location of high and low flow zones reflecting spatial terrain properties. The quartile and probability of exceedance maps can be used to illustrate attribute continuity. An added benefit is the identification of horizontal flux component q_h related to soil anisotropy. Distributions of q_h are strongly slope and aspect dependent and reflect contributions by the percolate flux q_v, and local anisotropy n. In this case, because of model simplicity, the approach was to simulate first and model later. Thus I have used conditionally simulated values of q_v and n at all nodes, to model distribution of q_h.

Correlation statistics of conditioning data per se are of little use in identifying spatially related zones, or similarity in behavior of data distributed in time. For example, although *SPAW*-modeled soil water distributions for the 101 conditioning data were correlated with recharge flux in the 4th and 6th layer, visual evidence to support it at a spatial scale was limited, and other factors such as slope, aspect, and soil type had to be also considered. On the other hand, *SPAW*-modeled soil water distributions in the 101 pedons showed good correlation with *ET* at the surface and only between adjoining layers with depth, while inspection of kriged θ_v overlays suggested that correlation with depth should also be present between other than adjoining layers.

Because variability of soil properties is so pervasive, application-oriented modeling of flow and transport at any scale requires that variability considerations be included in all estimation procedures. Applications envisage potential uses in precision farming, land use and management, alley farming, agroforestry, and nonpoint source and groundwater pollution control. Spatial soil properties can seldom be described in terms of simple statistics and may require application of more sophisticated analysis where their interdependence is a function of space, time, and association criteria. The uncertainty associated with modeling of any process arises because of our inability to completely characterize every aspect of the phenomena (Goovaerts, 1997). The accuracy of model prediction depends on the quality and adequacy of the data as well as the conceptual accuracy of the model. However, no model can ever truly validate an open, complex, natural system, since we seldom have adequate (exhaustive) data, and can never fully conceptualize the natural phenomena (Oreskes et al., 1994). Thus there is always a need to evaluate associated uncertainty that arises from sampling, and document the limitations of both the data and the model.

ACKNOWLEDGMENT

Contributions by D. Simmons to data visualization and processing are appreciated.

REFERENCES

Ahuja, L.R., O. Wendroth, and D.R. Nielsen. Relationship between initial drainage of surface soil and average profile saturated conductivity, *Soil Sci. Soc. Am. J.,* 57, pp. 19–25, 1993.

Ciolkosz, E.J. and N.C. Thurman. *Soil Char. Lab. Database System,* Agron. Ser. No. 124. Agron. Dept., Penn State Univ., University Park, PA, 1992, p. 59.

Cressie, N.A.C. *Statistics for Spatial Data.* John Wiley & Sons, Inc., New York, 1991, p. 900.

De Gruijter, J.J., D.J.J. Walvoort, and P.F.M. van Gaans. Continuous soil maps—A fuzzy set approach to bridge the gap between aggregation levels of process and distribution models, *Geoderma* 77, pp. 169–195, 1997.

Deutsch, C.V. and A.G. Journel. *GSLIB, Geostatistical Software Library and Users Guide,* Oxford University Press, New York, 1992, 1997, pp. 340, 369.

Eastman, J.R. *IDRISI: A Grid Based Geographic Analysis System,* Version 4.1, Clark University Graduate School of Geography, Worcester, MA, 1992, p. 78.

Gomez-Hernandez, J.J. and R.M. Srivastava. ISIM3D: An ansi-C three-dimensional multiple indicator conditional simulation program, *Comp. Geosci.* 16(4), pp. 395–440, 1990.

Goovaerts, P. *Geostatistics for Natural Resources Evaluation,* Oxford University Press, New York, 1997, p. 483.

Isaaks, H. and R.M. Srivastava. *Applied Geostatistics,* Oxford University Press, New York, 1989, p. 561.

Journel, A.G. *Fundamentals of Geostatistics in Five Lessons.* American Geophysical Union, Short Course in Geology: Volume 8, Washington, DC, 1989, p. 40.

Journel, A.G. A Return to the Equiprobability of Stochastic Realizations, in *Annual Report #8,* Stanford Center for Reservoir Forecasting, Stanford University, Stanford, CA, 1995.

Journel, A.G. Modeling uncertainty and spatial dependence: stochastic imaging, *Int. J. Geogr. Inf. Syst.* 10(5), pp. 517–522, 1996.

Journel, A.G. and F. Alabert. Non-Gaussian data expansion in earth sciences. *Terra Nova* 1, pp. 123–134, 1989.

Journel, A.G. and C.V. Deutsch. Entropy and spatial disorder, *Math. Geol.,* 25(3), pp. 329–355, 1993.

Kang, B.T. and G.F. Wilson. The Development of Alley Cropping as a Promising Agro-Forestry Technology, in *Agro-Forestry, a Decade of Developments,* H.A. Steppler and P.K.R. Nair, Eds., ICRAF, Nairobi, Kenya, 1987, pp. 227–243.

Olea, R.A. *Geostatistical Glossary and Multilingual Dictionary,* Oxford University Press, New York, 1991, p. 177.

Oreskes, N., K. Shrade-Frechette, and K. Belitz. Verification, validation, and confirmation of numerical models in the earth sciences, *Science* 263, pp. 641–646, 1994.

Rogowski, A.S. Quantifying the Model of Uncertainty and Risk Using Sequential Indicator Simulation, in *Data Variability and Risk Assessment in Soil Interpretations,* W.D. Nettleton, A.G. Hornsby, R.B. Brown, and T.L. Coleman, Eds., SSSA Special Publication No. 47. Soil Science Society of America, Madison, WI, 1996a.

Rogowski, A.S. GIS modeling of recharge on a watershed, *J. Environ. Qual.* 25, pp. 463–474, 1996b.

Rogowski, A.S. and J.K. Wolf. Incorporating variability into soil map unit delineations, *Soil Sci. Soc. Am. J.,* 58, pp. 163–174, 1994.

Rogowski, A.S. and J.R. Hoover. Catchment infiltration I: Distribution of variables, *Transactions in GIS* 1(2), pp. 95–110, 1996.

SAS. *SAS Procedures Guide,* Version 6, 3rd ed., SAS Institute, Inc., Cary, NC, 1995, p. 705.

Saxton, K.E. *SPAW (Soil-Plant-Atmosphere-Water) Model. Users Manual,* 3.3. USDA-ARS, Pullman, WA, 1994.

Saxton, K.E., W.J. Rawls, J.S. Romberger, and R.I. Papendick, Estimating generalized soil-water characteristics from texture. *Soil Sci. Soc. Am. J.,* 50, pp. 1031–1036, 1986.

Saxton, K.E., M.A. Porter, and T.A. McMahon. Climatic impacts on dryland winter wheat by daily soil water and crop stress simulations, *Agric. For. Meteorol.,* 58, pp. 177–192, 1992.

Sinowski, W., A.C. Scheinost, and K. Auerswald. Regionalization of soil water retention curves in a highly variable soilscape, II, Comparison of regionalization procedures using a pedotransfer function, *Geoderma* 78, pp. 145–159, 1997.

Srivastava, R.M. An overview of stochastic spatial simulation, in *Spatial Accuracy Assessment in Natural Resources and Environmental Sciences,* H.T. Mowrer, R.L. Czaplewski, and R.H. Hamre, Eds., Second International Symposium, USDA Forest Service, Fort Collins, CO, 1996, pp. 13–22.

Stein, A., I.G. Staritsky, J. Bouma, A.C. Van Eunsbergen, and A.K. Bregt. Simulation of moisture deficits and areal interpolation by universal cokriging, *Water Resource Res.,* 27(8), pp. 1963–1973, 1991.

Ulaby, F.T., R.K. Moore, and A.K. Fung. *Microwave Remote Sensing: Active and Passive,* Vol. II, Addison-Wesley, 1982, p. 861.

Van Schilfgaarde, J. Approximate solutions to drainage flow problems, in *Drainage of Agricultural Lands,* J.N. Luthin, Ed., Agronomy Monograph #7, American Society of Agronomy, Madison, WI, 1957.

White, R.A. *Earth Systems Science Center EOS Database,* Pennsylvania State University, http://eoswww.essc.psu.edu/eosdb.html, 1997.

Zaslavsky, D. and A.S. Rogowski. Hydrologic and morphologic implications of anisotropy and infiltration in soil profile development, *Soil Sci. Soc. Am. J.,* 33(4), pp. 594–599, 1969.

Chapter 16

Analyzing Spatiotemporal Dynamics of Plant Populations and Vegetation

Paul Braun, Heiko Balzter, and Wolfgang Köhler

INTRODUCTION

Much is known about temporal dynamics of plant populations and vegetation (Begon and Mortimer, 1986; Gillman and Hails, 1997). The interest in these dynamics is driven by various disciplines. Farmers, for example, wish to maximize yields, therefore enhancing population dynamics of crops (Hay and Walker, 1989). Conservation biologists, as another example, may need to estimate the fate of rare species and how differing management methods influence their population dynamics (Krebs, 1994; Meffe and Carroll, 1994). Finally, landscape planning requires knowledge about the temporal dynamics of plant populations in order to select appropriate species for revegetation projects (Jordan et al., 1987). Thus many ways to analyze and visualize temporal dynamics of plant populations and vegetation are at hand (Barbour et al., 1987; Greig-Smith, 1983). Less work has been done on the spatial structure of plant populations, however.

Relatively few efforts have been made to analyze spatiotemporal dynamics of plant populations or vegetation, although this is thought to be an important aspect of ecological processes (Bascompte and Solé, 1995; Czárán and Bartha, 1992). Doubt still exists as to whether it makes sense to analyze spatiotemporal dynamics of plants. One argument is that gross dynamics are not influenced substantially by including spatial heterogeneities in population dynamics, meaning that decision-makers can still rely on temporal population dynamics approaches (Roughgarden, 1998). Indeed, experimental approaches are complicated considerably if spatial aspects are included in the analysis of plant population dynamics. But, as Tilman and Kareiva (1998) pointed out, spatial perspectives can provide significant insights into numerous ecological questions. We believe, too, that spatial heterogeneity influences the outcome of ecological and evolutionary processes. An example will illustrate our reasoning: let three equally sized clusters of plants form a population in a defined area. If two of them become extinct and the third one increases slightly, the overall population dynamics would signal population decrease. This might be misleading, as the third cluster could be an exceptionally fit subpopulation which is increasing in size, or a cluster in which local resources abound. Any knowledge about the spatial distribution of a plant population allows conclusions to be drawn regarding ecologically relevant factors. This enables us to better understand the population dynamics of specific plant populations or vegetation, which in turn is important for plant invasive processes or plant extinctions.

The aim of this chapter is to outline methods that can be used by experimentators to tackle the problem of sampling and processing data on spatiotemporal plant population and vegetation dynamics. Our second aim is to show that spatiotemporal dynamics of plant populations yield more valuable information than investigating solely temporal population dynamics. This chapter presents two examples, where standard methods have been used to describe spatiotemporal dynamics of a plant population and a vegetation.

Material and Methods

There are two steps in the analysis of spatiotemporal dynamics of plant populations: sampling and data processing. When discrete objects/individual plants are identifiable, common methods of quantitative ecology are used to further process the data (Greig-Smith, 1983; Elliott, 1971; Kershaw, 1978; Upton and Fingleton, 1990; Vandermeer, 1981). These methods can be grouped as either aggregation or distance methods. Whereas aggregation methods count individuals (e.g., Variance to Mean Ratio, Morisita Index), distance methods (e.g., Nearest Neighbor Index, L-Plot) use interplant distances to conclude whether plant dispersion is random, regular, or clustered.

When plants cannot be assigned to a single point, as is the case with clonal plants or vegetation, the above-mentioned methods are less useful for describing spatial dynamics of populations. Therefore, vegetation science uses different sampling methods (e.g., Relevé-, Quadrat-, Point-Method) to quantify the proportions of species in a plant community where discrete objects cannot be counted (Barbour et al., 1987; Sutherland, 1996). The Point-Method provides data to which a Moving-Window Algorithm can be applied (Balzter et al., 1995).

The Spatiotemporal Dynamics of a Thistle Population (*Cirsium arvense* [L.] Scopoli)

After mapping plant populations on a recultivated rubbish dump near Giessen, Germany, in 1991, we used these data to examine various aggregation and distance methods to quantify the spatial distribution of plant populations (Braun and Lachnit, 1993; Braun and Lachnit, 1994). We found the Nearest Neighbor Index (R) to be a suitable statistic for describing spatial and temporal changes in plant populations of discrete objects. R depends on the distance of each plant to its nearest neighbor. As this information cannot be easily obtained under field conditions, we mapped the whole plant population (*C. arvense*) in a 2×2 m plot. With these coordinates, the average distances r_M of all individual plants to their nearest neighbors were calculated and normalized by an expected nearest neighbor distance, r_E. This is expressed by

$$R = r_M / r_E \tag{1}$$

where R is the ratio between the measured average nearest neighbor distance r_M and the expected nearest neighbor distance r_E (i.e., random distribution of individual plants in the population). The value of r_M is obtained by

$$r_M = (1/N) \cdot \sum_{i=1}^{N} r_i \tag{2}$$

where r_i is the distance of plant i to its nearest neighbor and N is the population size. The expected value r_E is defined as

$$r_E = 1/\left(2 \cdot \sqrt{u}\right), \quad \text{with } u = N/A \qquad (3)$$

Here A stands for the area of the mapped plot. R close to 1 indicates randomness, $R<1$ indicates a clustered, and $R>1$ indicates a regularly dispersed plant population. To test the significance of R, Clark and Evans (1954) proposed the following test statistic:

$$C = \left(r_M - r_E\right)/\sigma_E \qquad (4)$$

where

$$\sigma_E^2 = (4 - \pi)/(4 \cdot N \cdot u \cdot \pi) \qquad (5)$$

C can be tested with a standard normal distribution.

The Spatiotemporal Dynamics of a Meadow Vegetation (Lolio-Cynosuretum)

For continuous data; i.e., when plants are not recorded at single points, aggregation and distance methods are not efficient. We therefore sampled an area by the Point-Method and processed these raw data with a Moving Window Algorithm (MWA). The investigation was conducted on a meadow in Giessen, Germany. Except for the center, which remained unmowed, the meadow was cut up to 10 times a year. This led to an association of *Lolio-Cynosuretum* on the outer part of the meadow (Figure 16.1), whereas in the unmowed center the plant community could only be assigned the class of *Molinio-Arrhenatheretea* (Oberdorfer, 1990).

As shown in Figure 16.1, the study area was divided into 120 subplots. Inside each subplot a vegetation sample was taken with a point quadrat frame (three needles, ø 1 mm, each 30 cm apart). A vegetation sample consisted of all contacts of plant species with any of the three needles in one frame.

Four subplots or one window half were pooled to yield one multivariate vector, with each vector element representing one plant species. Each element of the vector took on the mean number of contacts which the respective plant species had with the three needles. To compare two successive years, vegetation data of the same window half, but for two successive years were used. These two window halves gave two multivariate vectors for which the Squared Euclidian Distance (SED) was calculated:

$$SED_{nw} = \sum_{i=1}^{s} \left(\overline{x}_{Ai} - \overline{x}_{Bi}\right)^2 \qquad (6)$$

Variables n and w denote the number and size of the window, respectively. The total species number is s. Index A stands for the window half of the first year and B for the window

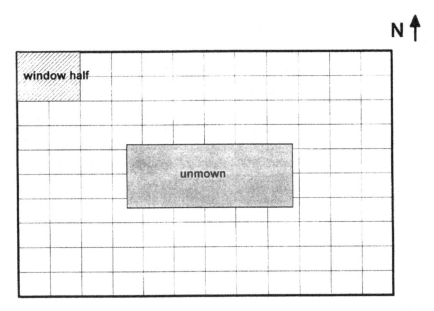

Figure 16.1. Investigated meadow in Giessen, Germany (630 m^2). The whole area is divided into 120 subplots, which were sampled every three months. The central part remained unmowed (72 m^2).

half of the second year. For every species the difference between years was calculated and squared. The sum of these values for all species resulted in one value, the Squared Euclidian Distance between the two multivariate vectors of successive years. This figure indicates how similar or how different the meadow was at this position between two years. An SED value of zero means that no phytomass change occurred, although compositional changes may have taken place. High SED values indicate a large vegetation change between the years. Note that this is a combined measure of species number and phytomass. In our example we moved the window stepwise, or subplotwise, from left to right (Figure 16.1). Thus the window stretched over two years and was moved in space. For one row this gave 11 values. After finishing a row the procedure continued with the next row. The overlap here was again one subplot. Thereby we could describe the vegetation change between years by 11×9 points, where each value represents a window comparing two years.

Results

The Spatiotemporal Dynamics of a Thistle Population (Cirsium arvense [L.] Scopoli)

The dynamics of the *R* values for a thistle population with the respective population dynamics (N) of *C. arvense* are shown as R/N-Plot in Figure 16.2. *R* was calculated for six different sampling dates in 1991. Looking at the population size (N) reveals abundance dynamics commonly found in plant populations of Central Europe. The population size in the 2×2 m plot was approximately the same at the beginning and at the end of the season (Figure 16.2). Viewing solely the abundance dynamics it is impossible to decide whether

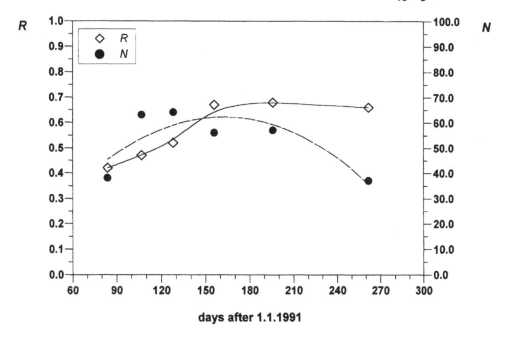

Figure 16.2. Spatial structure (Nearest Neighbor Index, R) and temporal dynamics (population size, N) of a thistle population (*Cirsium arvense*) in 1991. B-Spline curve fittings are shown for R (solid line) and for N (broken line). Sampling dates were: 25.3., 17.4., 8.5., 5.6., 15.7. and 19.9.1991.

the population is clustered or randomly dispersed. Furthermore, no change is detectable in the spatial composition of the population when analyzing temporal dynamics alone.

The results suggest a spatial dynamic in the population as indicated by the *R* value. Our data show a decrease in clustering of the thistle population during the course of the year (Figure 16.2). The question is whether clustering of the thistle population decreased in 1991 or not. To test this hypothesis a linear regression was applied. Linear regression of *R* against time was significantly different from zero. It thus seems that the thistle population spreads throughout the year, although it did not increase in number. We infer this from the fact that the spatial distribution of *C. arvense* was less aggregated on the last than on the first sampling date. Thus, thistle plants in clusters contributed less to the final population size than single standing plants. Population size and Nearest Neighbor Index were not significantly correlated. The observed loss in aggregation would not have been detected by looking at the population dynamics alone.

The Spatiotemporal Dynamics of a Meadow Vegetation (Lolio-cynosuretum)

While the population in the preceding section was countable and positions of individual plants could be marked, in the following example this is not the case. Here plants are no longer discrete objects, but continuous elements of vegetation. Raw data were obtained using the Point-Method and processed using MWA analysis. Figure 16.3 shows the vegetation change from 1993 to 1994. There were fluctuations in the vegetation, but no obvious

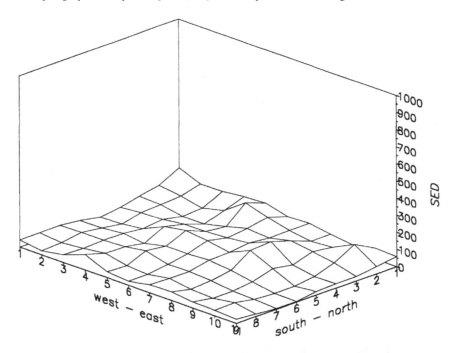

Figure 16.3. Plot of SED (Squared Euclidian Distance) between the vegetation at 14.5.1993 and 20.6.1994. An orchard is located north of the plot; a lane is to the west.

systematic change. A slight ridge formed, starting from the northeastern edge to the west-east side of the field at position 4/9 (Figure 16.3). This structure signals that the vegetation here changed more than in other parts of the meadow. Rarely did no vegetation shifts occur between years, but see positions 11/7 and 11/6 (Figure 16.3). In both cases either species composition plus species phytomass did not change, or changes in some species were balanced out by changes in other species.

To get an idea whether these spatiotemporal dynamics would be more or less the same over several years, we computed the SED values for 1994/95 (Figure 16.4). The change between the years 1994 and 1995 was pronounced. Most obviously, the vegetation in the unmowed central part of the meadow altered considerably (Figure 16.4). At the western side of the experimental field, slight vegetation changes between years took place, while at the eastern side, vegetation dynamics remained low. We conclude that considerable vegetational change occurred in the central part of the meadow. Even a shift of this size was not easy to detect by routine inspections of the field, but with these results in hand we again searched for and found changes in the meadow. The Point-Method plus MWA analysis has proved sensitive for detecting vegetational changes and the shift described here might easily have been overlooked without an MWA analysis.

Unlike the between-year shift of Figure 16.3, we assume that there were distinctions between locations on the meadow in the period from 1994 to 1995. A ridge structure as in Figure 16.3 is no longer visible. A comparison between the spatiotemporal vegetation dynamics of 1993/1994 and 1994/1995 shows that vegetation dynamics can vary greatly within the same area. We conclude that the vegetation under investigation is highly dynamic if viewed on a spatiotemporal scale.

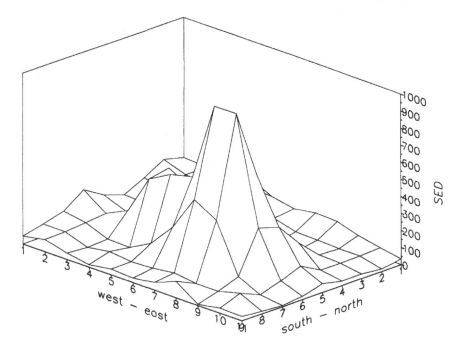

Figure 16.4. Plot of SED between the vegetation at 20.6.1994 and 16.6.1995. An orchard is located north of the plot; a lane is to the west.

DISCUSSION

The case studies presented above show a way to analyze and visualize spatiotemporal population and vegetation dynamics. They form a link between theory and data sampling. In both examples we discovered features that would have been overseen by analyzing the temporal dynamics of populations or vegetation alone.

From our experience with methods describing spatiotemporal dynamics of discrete plants, we conclude that there are few efficient methods to solve this task. As Ripley (1986) pointed out, most of the aggregation methods and some of the distance methods can be misleading in the interpretation of spatial structures. For a close inspection of a spatial structure, Ripley's L-Plot is recommended. The informative density of this statistic and our wish to obtain a quick impression of spatiotemporal dynamics in one figure led to our preference for the *R*-statistic. *R* is easily combined with population dynamics data (N), in a single graph which shows spatiotemporal dynamics (Figure 16.2). Even small changes in spatial structure are detectable.

The advantage of a combination of Point-Method and MWA is that it yields good estimates of even slight changes in vegetation, population, or phytomass. However, since SED figures are relative values, one can only compare the distance between specific window halves. In other words, adjacent SED's in Figure 16.3 or 16.4 may derive from different species compositions, even if they possess the same value. Another disadvantage of MWA was that we could not find a standard test of significance. Cornelius and Reynolds (1991) provide a permutation test for this purpose. However, compared to other procedures in vegetation analysis, like the Relevé-Method or the Quadrat-Method, the Point-Method with MWA better quantifies spatiotemporal vegetation shifts.

One problem still remains unsolved: the shape and position of structures in an investigation plot cannot be described by the above methods. This information might be necessary to judge specific population dynamics based on clustered resources. Thus we recommend always taking an additional look at the raw data. Another approach to this problem is the use of kriging methods (Isaaks and Srivastava, 1989). Kriging is a valuable, though unusual, tool for mapping population and vegetation distributions. Depending on the variable analyzed, contour maps can be drawn of population densities or vegetation phytomass. One must keep in mind, however, that contour maps result in a different type of information than the methods used in this chapter. R/N plots show the temporal changes in spatial distribution types of plant populations, MWA analysis allows comparisons between years at specific positions in the vegetation. Neither of these aspects is achieved by kriging methods. Furthermore, we assume that presentations of temporal dynamics with contour plots are difficult to interpret. However, we also believe that kriging methods are particularly helpful for the spatial presentation of plant populations and vegetation on a large geographical scale (Maurer, 1994). Nonetheless, kriging on a small scale could be applied more often than it is done at present.

We propose that the study of spatiotemporal dynamics of herbaceous plant populations should be intensified and methodological improvements made. Although a lot of theoretical work and modeling has already been done in this area (Tilman and Kareiva, 1998), little attention has been devoted to the measurement of spatiotemporal dynamics. Additional discussion is needed on the accurate measuring of spatiotemporal dynamics in plant communities.

REFERENCES

Balzter, H., P. Braun, and W. Köhler. Detection of Spatial Discontinuities in Vegetation Data by a Moving Window Algorithm, in *From Data to Knowledge,* W. Gaul and D. Pfeifer, Eds., Springer, Heidelberg, Germany, 1995.

Barbour, M.G., J.H. Burk, and W.D. Pitts. *Terrestrial Plant Ecology.* Benjamin/Cummings Publishing Company, Menlo Park, CA, 1987, p. 634.

Bascompte, J. and R.V. Solé. Rethinking complexity: modelling spatiotemporal dynamics in ecology, *Trends Ecol. Evol.* 10, pp. 361–366, 1995.

Begon, M. and M. Mortimer. *Population Ecology,* 2nd ed., Blackwell, Oxford, UK, 1986, p. 220.

Braun, P. and B. Lachnit. Characterizing the spatial distribution categories in plant populations, *Biometr. Bull.* 10, p. 25, 1993.

Braun, P. and B. Lachnit. Kennzeichnung der räumlichen Verteilung von Pflanzenpopulationen, *Zeitschrift für Agrarinformatik,* 4, pp. 67–71, 1994.

Clark, P.J. and F.E. Evans. Distance to nearest neighbor as a measure of spatial relationships in populations, *Ecology,* 35, pp. 445–453, 1954.

Cornelius, J.M. and J.F. Reynolds. On determining the statistical significance of discontinuities within ordered ecological data, *Ecology,* 72, pp. 2057–2070, 1991.

Czárán, T. and S. Bartha. Spatiotemporal dynamic models of plant populations and communities, *Trends Ecol. Evol.* 7, pp. 38–42, 1992.

Elliott, J.M. *Some Methods for the Statistical Analysis of Samples of Benthic Invertebrates,* Freshwater Biological Association, Ambleside, UK, 1971, p. 144.

Gillman, M. and R. Hails. *An Introduction to Ecological Modelling,* Blackwell, Oxford, 1997, p. 202.

Greig-Smith, P. *Quantitative Plant Ecology,* Blackwell, Oxford, UK, 1983, p. 359.

Hay, K.M. and A.J. Walker. *Crop Yield,* Longman, Harlow, England, 1989, p. 292.

Isaaks, E.H. and R.M. Srivastava. *Applied Geostatistics,* Oxford University Press, New York, 1989, p. 561.

Jordan, W.R., M.E. Gilpin, and J.D. Aber. *Restoration Ecology,* Cambridge University Press, Cambridge, UK, 1987, p. 342.

Kershaw, K.A. *Quantitative and Dynamic Plant Ecology,* Edward Arnold, London, UK, 1978, p. 303.

Krebs, C.J. *Ecology,* HarperCollins College Publishers, New York, 1994, p. 801.

Maurer, B.A. *Geographical Population Analysis: Tools for the Analysis of Biodiversity,* Blackwell, Oxford, UK, 1994, p. 130.

Meffe, G.K. and C.R. Carroll. *Principles of Conservation Biology,* Sinauer Associates, Sunderland, MA, 1994, p. 600.

Oberdorfer, E. *Pflanzensoziologische Exkursionsflora,* Ulmer, Stuttgart, Germany, 1990, p. 1050.

Ripley, B.D. Spatial Point Pattern Analysis in Ecology, in *Developments in Numerical Ecology,* P. Legendre and L. Legendre, Eds., Springer, Berlin, Germany, 1986.

Roughgarden, J. Production Functions from Ecological Populations: A Survey with Emphasis on Spatially Implicit Models, in *Spatial Ecology,* D. Tilman and P. Kareiva, Eds., Princeton University Press, Princeton, NJ, 1998.

Sutherland, W.J. *Ecological Census Techniques,* Cambridge University Press, Cambridge, UK, 1996, p. 336.

Tilman, D. and P. Kareiva. *Spatial Ecology,* Princeton University Press, Princeton, NJ, 1998, p. 368.

Upton, G.J.G. and B. Fingleton. *Spatial Data Analysis by Example,* John Wiley, Chichester/New York, 1990, p. 410.

Vandermeer, J. *Elementary Mathematical Ecology,* John Wiley, New York, 1981, p. 294.

Chapter 17

Space-Time Statistical Modeling of Environmental Data

Christopher K. Wikle and Noel Cressie

INTRODUCTION

As a concept, the environment is simply the surroundings of an organism or organisms. Space and time scales in environmental investigations can range from the very local to the very global. Some studies attempt to understand physical, chemical, and biological processes by performing controlled experiments in the laboratory. We are concerned instead with studies made in the field. These are observational in nature and, hence, even though a large amount of data may be collected and analyzed, one typically can only infer associations rather than causation.

Most environmental studies in the field involve variability over both space and time. The extension of traditional geostatistical methods, such as kriging, to the space-time domain is one possible approach to characterizing the variability of such processes (e.g., Bilonick, 1983; Eynon and Switzer, 1983; Stein, 1986; Le and Petkau, 1988; Sampson and Guttorp, 1992; Handcock and Wallis, 1994; Host et al., 1995). However, there are difficult modeling decisions to make in this approach, involving space, time, and space-time interaction components. The space-time variability can be characterized by a variogram that often exhibits very different spatial behavior at different points in time and the class of variogram models that can be fit in this situation is very small indeed. In the atmospheric sciences, traditional methods for examining space-time processes have focused on Empirical Orthogonal Functions (EOF), Canonical Correlation Analysis (CCA), and Principal Oscillation Patterns (POP); see von Storch and Zwiers (1999) for an overview of these methods. Although these techniques provide powerful ways to summarize the huge space-time data sets common in atmospheric and related sciences, they were not designed with space-time prediction in mind.

Without the spatial component, there is a large class of time series that could be used to model the temporal component (e.g., autoregressive error processes). These are *dynamic* in the sense that they exploit the unidirectional flow of time. Without the temporal component, geostatistical methods could be used to model the spatial component (e.g., intrinsically stationary error processes). These are *descriptive* in the sense that although they model spatial correlation there is no causative interpretation associated with them. When both temporal and spatial components are present, it seems sensible to use models that are a combination of both approaches, namely temporally dynamic *and* spatially descriptive. That is the new feature of our work and it allows a natural development of the space-time Kalman filter (Wikle, 1996; Wikle and Cressie, 1997).

State-space Kalman filter (Kalman, 1960) approaches have been considered in the atmospheric and oceanic sciences since Ghil et al.'s (1981) pioneering work. (For a recent overview, see Cohn, 1997, and Ghil, 1997.) These approaches have generally been developed in the context of Numerical Weather Prediction (NWP) data initialization. In NWP applications, it is assumed that the true state of the atmosphere (or ocean) evolves according to a physical, albeit simplified, multivariate model. Practically, these Kalman filters have been limited by the tremendous computational burden associated with state processes that have on the order of 10^6 variables. There have been recent attempts to reduce the dimensionality of the Kalman filters associated with these physical models (e.g., Fukumori and Malanotte-Rizzoli, 1995). Although these Kalman filter methods have great potential for optimally preparing atmospheric data streams to be used by physical numerical prediction models, the physically based state matrices prevent these approaches from being used for spatiotemporal processes where explicit physical models are not well understood (e.g., precipitation). Thus, there is a need for space-time Kalman filters of dynamical processes that are based on statistically derived state matrices.

In a published comment to Handcock and Wallis's (1994) space-time approach, Cressie (1994) suggested that a Kalman filter incorporating space and time would be a powerful way to apply the Bayesian paradigm to spatiotemporal problems. Subsequently, Huang and Cressie (1996) developed a prototypical temporally dynamic and spatially descriptive model in a Kalman filter framework. Their model assumes that the prediction at some location **s** and time t is influenced directly by past values only at location **s**. Environmental spatiotemporal processes are likely to be more complicated and may also show dependence on past values at locations near **s**. Wikle (1996) and Wikle and Cressie (1997) extended the model of Huang and Cressie (1996) to include dynamical contributions from all locations in the spatial domain of interest. This model can incorporate relatively complicated space-time dynamics, allows for nonparametric, nonstationary, and anisotropic covariance structures, can predict at locations at which data have never been observed (in space and/or time), works well in low signal-to-noise-ratio environments, and is implemented via a truncated set of orthogonal basis functions in a Kalman-filter setting. This implementation allows the Kalman filter to be used efficiently in large-data-set environments. A similar, yet independent, implementation (referred to as a *reduced-state-space* Kalman filter) was reported in the atmospheric science literature by Cane et al. (1996). The principal differences between our approach and the Cane et al. approach is that they cannot make predictions for locations at which observations have never been reported, and our formulation naturally leads to an additional kriging-like spatial model for the nondynamic portion of the space-time process.

Our goal in this chapter is to demonstrate to researchers in the environmental sciences that our space-time Kalman filter technology can be used to solve difficult problems in space-time prediction. To do this, we borrow some of the technical description from Wikle and Cressie (1997), although in an abbreviated format. However, our emphasis here is on the application of this methodology (prediction of precipitation in the South China Sea), rather than its technical exposition. The remainder of this chapter is organized as follows. The next section gives an overview of our model and the Kalman filter implementation. A brief description of parameter estimation is included in the following section. Application of our methodology to precipitation in the South China Sea is given next. Climate prediction, and in particular precipitation prediction, is a vexing environmental problem and for this reason, we feature it in this chapter. Finally, a brief discussion of extensions to our approach is presented in the final section.

MODEL DEVELOPMENT

A complete development of the model can be found in Wikle (1996) and Wikle and Cressie (1997). Some of the basic features are repeated here for clarity of presentation. Consider a spatially continuous process $Z(\mathbf{s};t)$, where $\mathbf{s} \in D$, with D some spatial domain in d-dimensional Euclidean space R^d, and a discrete index of times $t \in \{1,2,...\}$. We suppose that the observable process has a component of measurement error expressed through the measurement equation,

$$Z(\mathbf{s};t) = Y(\mathbf{s};t) + \varepsilon(\mathbf{s};t) \tag{1}$$

where $Y(\mathbf{s};t)$ can be thought of as a "smoother" process than $Z(\mathbf{s};t)$. Our goal is to predict $\{Y(\mathbf{s};t) : \mathbf{s} \in D\}$ from data $\{Z(\mathbf{s}_1;\tau),...,Z(\mathbf{s}_n;\tau) : \tau=1,...,t\}$. Now, we assume that $Y(\mathbf{s};t)$ has zero mean and can be written,

$$Y(\mathbf{s};t) = Y_K(\mathbf{s};t) + v(\mathbf{s};t) \tag{2}$$

where $v(\mathbf{s};t)$ is a component of variance representing small-scale spatial variation that does not have a coherent temporally dynamic structure. The component $Y_K(\mathbf{s};t)$ is assumed to evolve according to the state equation,

$$Y_K(\mathbf{s};t) = \int_D w_s(\mathbf{u})Y_K(\mathbf{u};t-1)d\mathbf{u} + \eta(\mathbf{s};t) \tag{3}$$

where $\eta(\mathbf{s};t)$ is a spatially colored error process (i.e., the "spatially descriptive" component) and $w_s(\mathbf{u})$ is a function representing the interaction between the state process Y_K at location $\mathbf{u}$ and time (t–1) and Y_K at location $\mathbf{s}$ and time t (i.e., the "temporally dynamic" component). For stationarity over time, we further require that this interaction function satisfies

$$\int_D w_s(\mathbf{u})d\mathbf{u} = \alpha \tag{4}$$

where $|\alpha|<1$ is an unknown parameter. We also assume that the various error components in the model are uncorrelated.

The state model (2), (3) is an extension of the model given by Huang and Cressie (1996), who effectively assume that $v(\cdot;t) \equiv 0$ and that $w_s(\mathbf{u})$ is an unknown constant times the Dirac delta function, thereby only considering contributions to $Y_K(\mathbf{s};t)$ from previous values of the process at the *same* location $\mathbf{s}$. The strength of the state model (3) is that it features the dynamic aspect through the continuous autoregressive term but builds in spatiotemporal interaction through the error process $\eta(\cdot;t)$, which is, at any point in time, a spatially correlated (e.g., intrinsically stationary) process.

We are interested in the optimal predictor of $Y(\mathbf{s};t)$ given the data $\mathbf{Z}(t),...,\mathbf{Z}(1)$, where

$$\mathbf{Z}(\tau) \equiv [Z(\mathbf{s}_1;\tau), Z(\mathbf{s}_2;\tau),...,Z(\mathbf{s}_n;\tau)]'; \quad \tau = \{1,2,...,t\} \tag{5}$$

and $\mathbf{s}$ is a location at which we may or may not have observations. That is, we wish to obtain the predictor,

$$\hat{Y}(\mathbf{s};t|t) \equiv \mathrm{E}[Y(\mathbf{s};t)|\mathbf{Z}(t),...,\mathbf{Z}(1)] \tag{6}$$

and its mean-squared prediction error

$$\mathrm{E}[\hat{Y}(\mathbf{s};t|t) - Y(\mathbf{s};t)]^2 \tag{7}$$

In principle, both of these quantities can be calculated recursively using a *space-time Kalman filter* (STKF).

Reduced State-Space Representation

Instead of using the Kalman filter on the Y process as discussed above, we expand the Y_K process in terms of a set of complete and orthonormal basis functions, and use the associated expansion coefficients as the "state" process in the Kalman filter and extend the model's applicability to large-data problems. By utilizing this approach, we are able to reduce substantially the dimensionality of our state process in the Kalman filter. Then, the optimal predictor is a sum of two terms, one involving a space-time Kalman filter of the Y_K process and the other one analogous to optimal spatial prediction (i.e., simple kriging).

For large systems, the full state-space formulation given by (1), (2), and (3) is difficult to implement. Thus, we seek to reformulate the dynamical component in a state space of reduced dimension. We do this by expanding Y_K and the weight functions $w_s(\mathbf{u})$ in terms of a truncated set of complete and orthonormal basis functions, $\{\phi_k(\mathbf{u}) : k=1,...,\infty\}$:

$$Y_K(\mathbf{u};t) = \sum_{k=1}^{K} a_k(t)\phi_k(\mathbf{u}) \tag{8}$$

$$w_s(\mathbf{u}) = \sum_{l=1}^{K} b_l(\mathbf{s})\phi_l(\mathbf{u}) \tag{9}$$

where $a_k(t)$ are zero mean random variables and $b_l(\mathbf{s})$ are unknown nonstochastic parameters. Then, using (8) and (9) in (1)–(3), and making assumptions of uncorrelated errors, we obtain the reduced state-space model (see Wikle, 1996; Wikle and Cressie, 1997):

$$Z(\mathbf{s};t) = \phi(\mathbf{s})'\mathbf{a}(t) + v(\mathbf{s};t) + \varepsilon(\mathbf{s};t) \tag{10}$$

$$\mathbf{a}(t) = \mathbf{H}\mathbf{a}(t-1) + \mathbf{J}\eta(t) \tag{11}$$

where $\phi(\mathbf{s}) \equiv [\phi_1(\mathbf{s}),...,\phi_K(\mathbf{s})]'$ is a $K\times 1$ vector, $\mathbf{a}(t) \equiv [a_1(t),...,a_K(t)]'$ is a $K\times 1$ state vector, $\mathbf{H}$ is a $K\times K$ function of the first K basis vectors and the unknown parameters $\mathbf{b}(\mathbf{s}) \equiv [b_1(\mathbf{s}),...,b_K(\mathbf{s})]'$ at all measurement locations, $\mathbf{J} \equiv [\Phi'\Phi]^{-1}\Phi'$ is a $K\times n$ matrix that is a function of the first K basis vectors at each measurement location, $\Phi \equiv [\phi(\mathbf{s}_1),\phi(\mathbf{s}_2),...,\phi(\mathbf{s}_n)]'$, and $\eta(t) \equiv [\eta(\mathbf{s}_1;t),...,\eta(\mathbf{s}_n;t)]'$ is the $n\times 1$ error vector. Assuming we have an n-dimensional state process to begin with, we are now able to model it via (11), as a $K<<n$ dimensional process, with minimal loss of information. The choice of K can be based on the percentage of variance accounted for by the first K basis vectors. Of course, the choice of K must also consider the desired degree of reduction in dimensionality of the state process. Our experience suggests that many spatiotemporal environmental processes can be modeled successfully by choosing a K that accounts for 80–95% of the process variance (e.g., see Precipitation over the South China Sea).

IMPLEMENTATION

The reduced state-space model (10) and (11) can be implemented easily via a Kalman filter. Such an implementation allows updateable prediction of $\mathbf{a}(t)$ and the model's mean-squared prediction error, respectively:

$$\hat{\mathbf{a}}(t|t) \equiv E[\mathbf{a}(t)|\mathbf{Z}(t),...,\mathbf{Z}(1)] \tag{12}$$

$$= \hat{\mathbf{a}}(t|t-1) + \mathbf{K}(t)[\mathbf{Z}(t) - \Phi\hat{\mathbf{a}}(t|t-1)] \tag{13}$$

$$\mathbf{P}(t|t) \equiv E\{[\mathbf{a}(t) - \hat{\mathbf{a}}(t|t)][\mathbf{a}(t) - \hat{\mathbf{a}}(t|t)]' \tag{14}$$

$$= \mathbf{P}(t|t-1) - \mathbf{K}(t)\Phi\mathbf{P}(t|t-1) \tag{15}$$

where the Kalman gain $\mathbf{K}(t)$ is given by

$$\mathbf{K}(t) = \mathbf{P}(t|t-1)\Phi'[\mathbf{R} + \mathbf{V} + \Phi\mathbf{P}(t|t-1)\Phi']^{-1} \tag{16}$$

and the one-step-ahead predictions are defined as:

$$\hat{\mathbf{a}}(t|t-1) \equiv E[\mathbf{a}(t)|\mathbf{Z}(t-1),...,\mathbf{Z}(1)] \tag{17}$$

$$= \mathbf{H}\hat{\mathbf{a}}(t-1|t-1) \tag{18}$$

$$\mathbf{P}(t|t-1) \equiv var[\mathbf{a}(t)|\mathbf{Z}(t-1),...,\mathbf{Z}(1)] \tag{19}$$

$$= \mathbf{H}\mathbf{P}(t-1|t-1)\mathbf{H}' + \mathbf{J}\mathbf{Q}\mathbf{J}' \tag{20}$$

where $\mathbf{R} \equiv \mathrm{var}[\varepsilon(t)]$,

$\quad\quad \mathbf{V} \equiv \mathrm{var}[\mathrm{v}(t)]$,

$\quad\quad \mathbf{Q} \equiv \mathrm{var}[\eta(t)]$,

$\quad\quad \varepsilon(\mathrm{t}) \equiv [\varepsilon(\mathbf{s}_1;t),...,\varepsilon(\mathbf{s}_n;t)]'$, and

$\quad\quad \mathrm{v}(t) \equiv [\mathrm{v}(\mathbf{s}_1;t),...,\mathrm{v}(\mathbf{s}_n;t)]'$.

Given estimates $\hat{\mathbf{a}}(t|t)$, we can then obtain the estimates for the Y-process given the data and information concerning the covariance of the Z and v processes:

$$\hat{Y}(\mathbf{s};t|t) = \mathrm{E}[Y(\mathbf{s};t)|\mathbf{Z}(t),\mathbf{Z}(t-1),...,\mathbf{Z}(1)] \tag{21}$$

$$= \phi(\mathbf{s})'\hat{\mathbf{a}}(t|t) + \mathbf{c}_v(\mathbf{s})'\left[\mathbf{C}_0^Z\right]^{-1}\mathbf{Z}(t) \tag{22}$$

where $\mathbf{c}_v(\mathbf{s}) \equiv [c_v(\mathbf{s},\mathbf{s}_1),...,c_v(\mathbf{s},\mathbf{s}_n)]'$,

$\quad\quad c_v(\mathbf{s},\mathbf{r}) \equiv \mathrm{E}[\mathrm{v}(\mathbf{s};t)\mathrm{v}(\mathbf{r};t)]$, and

$\quad\quad \mathbf{C}_0^Z \equiv \mathrm{cov}[\mathbf{Z}(t),\mathbf{Z}(t)]$.

The first term in (22) relates to the prediction of the dynamic component of Y (i.e., Y_K). The second term in (22) concerns the prediction of the nondynamic component $\mathrm{v}(\mathbf{s};t)$, analogous to optimal interpolation (or simple kriging). In fact, a strength of our formulation is that as the truncation K becomes smaller, the solution approaches the optimal interpolation solution applied to spatial fields independent of time.

We can also derive the mean-squared prediction error for the Y-process:

$$\mathrm{var}[Y(\mathbf{s};t) - \hat{Y}(\mathbf{s};t|t)] = \mathrm{var}[Y(\mathbf{s};t)|\mathbf{Z}(t),\mathbf{Z}(t-1),...,\mathbf{Z}(1)]$$

$$= \phi(\mathbf{s})'\mathbf{P}(t|t)\phi(\mathbf{s}) + c_v(\mathbf{s},\mathbf{s}) - \mathbf{c}_v(\mathbf{s})'\left[\mathbf{C}_0^Z\right]^{-1}\mathbf{c}_v(\mathbf{s}) \tag{23}$$

$$-2\phi(\mathbf{s})'\mathrm{cov}[\hat{\mathbf{a}}(t|t),\mathbf{Z}(t)]\left[\mathbf{C}_0^Z\right]^{-1}\mathbf{c}_v(\mathbf{s}) \tag{24}$$

which is composed of variance-covariance terms related to the dynamic component, the nondynamical component, and their interaction. Through the Kalman filter, we can easily obtain predictions at times outside the data window, for either data or nondata locations (e.g., Wikle, 1996).

Selection of Basis Functions

As stated previously, we can choose any set of basis functions $\{\phi_k(\cdot)\}$, as long as they are complete, orthonormal, and are defined at any location $\mathbf{s}$ in domain D. Thus, there are many possible choices for these functions. For instance, we could choose eigenfunctions based on some *a priori* chosen physical (deterministic) model of the system of interest. Or, if we know little about the governing dynamics of the process of interest, we could choose some general basis set, either empirical (e.g., EOFs) or specified (e.g., orthogonal polynomials, wavelets). Theoretically, it makes no difference which approach we take. However, certain choices are more advantageous from a practical standpoint. As discussed below, we focus our attention here on the EOF basis set.

In two spatial dimensions, we choose the EOF basis because it has a long history of use in the atmospheric sciences and, more importantly, because it has certain optimality properties with regard to truncation; see below. The use of EOFs in spatial prediction has been considered by Cohen and Jones (1969), Creutin and Obled (1982), Obled and Creutin (1986), Shriver and O'Brien (1995), Smith et al. (1996), and Cane et al. (1996). Creutin and Obled (1982) point out that the EOF approach to spatial prediction naturally accounts for anisotropic and heterogeneous covariance structure. (See Guttorp and Sampson [1994] for a recent discussion of EOFs and spatial prediction.) A comprehensive overview of EOFs, related to their use as a summarization tool in diagnostic studies of large atmospheric data sets, can be found in Preisendorfer (1988).

When assuming a spatially continuous observation and state process, we consider the EOF basis through a Karhunen-Loève (K-L) expansion (e.g., see Papoulis, 1965, pp. 457–461). Given some spatiotemporal process $X(\mathbf{s};t)$ with $\mathbf{s} \in D$, $t \in \{1,2,...\}$, suppose that

$$E[X(\mathbf{s};t)] = 0 \tag{25}$$

and define the covariance function as

$$E[X(\mathbf{s};t)X(\mathbf{r};t)] \equiv c_0^X(\mathbf{s},\mathbf{r}) \tag{26}$$

which need not be stationary in space but is assumed to be invariant in time. The K-L expansion allows the covariance function to be decomposed as follows:

$$c_0^X(\mathbf{s},\mathbf{r}) = \sum_{k=1}^{\infty} \lambda_k \psi_k(\mathbf{s})\psi_k(\mathbf{r}) \tag{27}$$

where $\{\psi_k(\cdot) : k=1,...,\infty\}$ are the eigenfunctions and $\{\lambda_k(\cdot) : k=1,...,\infty\}$ are the associated eigenvalues of the Fredholm integral equation,

$$\int_D c_0^X(\mathbf{s},\mathbf{r})\psi_k(\mathbf{s})d\mathbf{s} = \lambda_k \psi_k(\mathbf{r}) \tag{28}$$

and

$$\int_D \psi_k(\mathbf{s})\psi_l(\mathbf{s})d\mathbf{s} = \begin{cases} 1 & \text{for } k = l \\ 0 & \text{otherwise} \end{cases} \tag{29}$$

Assuming completeness, we can then expand $X(\mathbf{s};t)$ according to

$$X(\mathbf{s};t) = \sum_{k=1}^{\infty} f_k(t)\psi_k(\mathbf{s}) \tag{30}$$

where we call $\{\psi_k(\mathbf{s}) : \mathbf{s} \in D\}$ the k-th EOF and often refer to the associated time series $f_k(t)$ as the k-th principal component time series, or "amplitude" time series. If we truncate the expansion (30) at K,

$$X_K(\mathbf{s};t) \equiv \sum_{k=1}^{K} f_k(t)\psi_k(\mathbf{s}) \tag{31}$$

then it can be shown (e.g., Freiberger and Grenander, 1965; Davis, 1976) that the EOF decomposition minimizes the variance of the truncation error, $E\{[X(\mathbf{s};t) - X_K(\mathbf{s};t)]^2\}$, and is thus optimal in this regard when compared to all other basis sets.

Data are finite and often irregularly located in space. Therefore, in practice we must solve numerically the Fredholm integral equation (28) to obtain the EOF basis functions. Cohen and Jones (1969) and Buell (1972, 1975) give numerical quadrature solutions to this problem. The numerical quadrature approaches for discretizing the integral equation succeed in that they give estimates for the eigenfunctions and eigenvalues that are weighted in some manner according to the spatial distribution of the data locations, but only for the eigenfunctions at locations $\{\mathbf{s}_1,...,\mathbf{s}_n\}$, for which there are data. Obled and Creutin (1986) provide an elegant extension of this quadrature approach which uses spline functions to interpolate the eigenfunctions and perform the quadrature simultaneously, similar to the "finite element" approach to numerical integration. An implementation of the Obled and Creutin (1986) approach in the current setting can be found in Wikle (1996).

As discussed by Buell (1972), when data are evenly distributed in space (e.g., gridded), EOF analysis is essentially equivalent to principal component analysis as defined in multivariate statistics. Thus, an alternative method of generating the EOF basis functions at all spatial locations of interest is by a *pregridding* procedure. That is, we use some simple spatial prediction scheme to predict at regular grid locations, whether we have data for that location or not. Then, we calculate the EOFs based on these "predicted" fields (e.g., Karl et al., 1982). For example, let the pregridded field be denoted by $\hat{Y}_{pg}(\mathbf{r}_i, t)$ for $i=1,...,m$, where $\{\mathbf{r}_i : i=1,...,m\}$ are the regular grid locations at which the predictions were made. Then, the EOF basis functions are obtained from the symmetric decomposition:

$$\hat{\mathbf{C}}_0^{Y_{pg}} = \boldsymbol{\Psi}\boldsymbol{\Lambda}\boldsymbol{\Psi}' \tag{32}$$

where $\hat{\mathbf{C}}_0^{Y_{pg}}$ is an estimate of the covariance matrix $\mathbf{C}_0^{Y_{pg}} \equiv E[\hat{\mathbf{Y}}_{pg}(t)\hat{\mathbf{Y}}_{pg}(t)']$, $\boldsymbol{\Lambda}$ is the diagonal matrix of eigenvalues from the spectral decomposition, and the EOFs are the columns of the eigenmatrix $\boldsymbol{\Psi} \equiv (\psi_1,...,\psi_m)$. To improve the empirical covariance estimation that is needed for determination of the EOF basis functions, we also smooth in time. There are any number of spatial prediction and time smoothing schemes that could be used for this purpose. In our example, we use a biharmonic spline procedure (Sandwell, 1987) as implemented in the MATLAB™[1] matrix language (MathWorks, 1996; referred to as an "inverse distance" gridding routine), as well as a Gaussian kernel smooth in time (e.g., Hastie and

---

[1] Registered trademark of the MathWorks, Inc., Natick, Massachusetts.

Tibshirani, 1990). For more details concerning this pregridding procedure, see Wikle and Cressie (1997).

ESTIMATION OF MODEL PARAMETERS

The Kalman filter presented in the previous section gives optimal predictors only if we know the true error covariances $\mathbf{R}$, $\mathbf{V}$, and $\mathbf{Q}$, as well as the state matrix $\mathbf{H}$ (note that $\mathbf{H} \equiv \mathbf{JB}$, where $\mathbf{B} \equiv [\mathbf{b}(\mathbf{s}_1),...,\mathbf{b}(\mathbf{s}_n)]$ and $\mathbf{J}$ is a known function of the basis vectors defined earlier). As described earlier, we can *choose* the ϕ's, as long as they are complete and orthonormal, but we must *estimate* the $\mathbf{R}$, $\mathbf{V}$, $\mathbf{Q}$, and $\mathbf{B}$ matrices. Although we no longer obtain the conditional expectation, but rather an estimate of it, our approach is completely analogous to kriging, where the variogram parameters are estimated. In both cases, we obtain a prediction method that is more broadly applicable to different physical processes because we let the data determine the structure of these matrices. In statistical terminology, this corresponds to viewing the Kalman filter as an empirical Bayesian technique.

We focus on simple method of moments (MOM) estimators for $\mathbf{R}$, $\mathbf{V}$, $\mathbf{Q}$, and $\mathbf{B}$. Although maximum likelihood (ML) estimators of model parameters are more efficient than MOM estimators, the high-dimensional nature of spatiotemporal problems makes for poorly behaved likelihood surfaces and iterative ML solutions that are difficult to implement. More detailed discussions concerning the estimation of model parameters can be found in Wikle (1996) and Wikle and Cressie (1997).

ESTIMATION OF MODEL COVARIANCES

This section describes the estimation of the $\mathbf{R}$, $\mathbf{V}$, and $\mathbf{Q}$ error covariance matrices, as defined above. Assuming the measurement error ε in (1) is spatial white noise (with variance σ_ε^2 we estimate $\mathbf{R}$ via:

$$\hat{\mathbf{R}} = \hat{\sigma}_\varepsilon^2 \mathbf{I} \tag{33}$$

where $\mathbf{I}$ is the $n{\times}n$ identity matrix and $\hat{\sigma}_\varepsilon^2$ is obtained from information about the measuring instrument, or through the behavior of an empirical variogram estimate of the data $\{\mathbf{Z}(t),...,\mathbf{Z}(1)\}$ as the spatial lag approaches zero. (For a discussion on variograms, their definition, estimation, and modeling, see Cressie [1993, Section 2.4].)

In addition to the estimate of $\mathbf{V}$ needed for the Kalman filter, (22) and (24) show that we may need an estimate of $\mathbf{c}_v(\mathbf{s})$, where $\mathbf{s}$ may be at a location where we do not have data. This suggests that we should *model* the covariance structure of the v process in (2).

We first obtain the spectral decomposition of

$$\mathbf{C}_0^Y \equiv \mathrm{cov}[\mathbf{Y}(t), \mathbf{Y}(t)] \tag{34}$$

given the truncated set of basis functions at data locations, $\mathbf{\Phi}$:

$$\mathbf{L} = \mathbf{J}\mathbf{C}_0^Y \mathbf{J}' \tag{35}$$

Transforming from the spectral domain back to physical space gives the covariance accounted for by the truncated basis set which is, by definition, the covariance matrix of the Y_K process at observation locations. That is,

$$\mathbf{C}_0^{Y_K} \equiv \mathbf{\Phi L \Phi}' \tag{36}$$

Then, the residual covariance matrix associated with the ν process is given by

$$\mathbf{V} = \mathbf{C}_0^Y - \mathbf{C}_0^{Y_K} \tag{37}$$

In practice, given an estimate of $\mathbf{C}_0^Y$, we can obtain an estimate of $\mathbf{V}$, which we denote $\hat{\mathbf{V}}$. In that case, care must be taken to assure the positive definiteness of $\hat{\mathbf{V}}$.

From $\hat{\mathbf{V}}$ we are only able to obtain estimates of the ν-process covariances at locations for which we have data. In order to obtain estimates at locations where we do not have data, we model $c_\nu(\mathbf{s}, \mathbf{r})$. This modeling is accomplished through some valid (i.e., positive-definite) covariance function (e.g., Cressie, 1993, pp. 84–86). In the case where $\hat{\mathbf{V}}$ is a diagonal matrix (or possibly diagonally dominant), we can assume that ν is simply white noise,

$$c_\nu(\mathbf{s}, \mathbf{r}) = \sigma_\nu^2; \ \mathbf{s} = \mathbf{r} \tag{38}$$

and zero otherwise. In that case, we estimate σ_ν^2 according to

$$\hat{\sigma}_\nu^2 = (1/n) \sum_{i=1}^n \hat{\nu}(\mathbf{s}_i, \mathbf{s}_i) \tag{39}$$

where $\hat{\nu}(\mathbf{s}_i, \mathbf{s}_i)$ is the i-th diagonal element of $\hat{\mathbf{V}}$. Typically, if the truncation parameter K is large, then a white noise assumption for $\hat{\nu}$ is plausible. See Wikle (1996) for a discussion of more general models than (38).

Examination of the Kalman filter equations (13)–(20) shows that we need only estimate $\mathbf{JQJ}'$, rather than $\mathbf{Q}$. It can be shown (Wikle, 1996) that:

$$\begin{aligned}
\mathbf{JQJ}' = \ &\mathbf{J}\left[\mathbf{C}_0^Z - \mathbf{V} - \mathbf{R}\right]\mathbf{J}' - \mathbf{JC}_1^Z\mathbf{J}'\mathbf{H}' \\
&-\mathbf{HJ}\left[\mathbf{C}_1^Z\right]'\mathbf{J}' + \mathbf{HJ}\left[\mathbf{C}_0^Z - \mathbf{V} - \mathbf{R}\right]\mathbf{J}'\mathbf{H}'
\end{aligned} \tag{40}$$

where

$$\mathbf{C}_1^Z \equiv \mathrm{cov}[\mathbf{Z}(t), \mathbf{Z}(t-1)] \tag{41}$$

and where we have assumed temporal invariance (e.g., lag-zero and lag-one temporal covariances do not depend on t).

Thus, to obtain the estimate $\mathbf{J}\hat{\mathbf{Q}}\mathbf{J}'$, we can substitute estimates for $\mathbf{C}_0^Z, \mathbf{C}_1^Z, \mathbf{V}, \mathbf{R},$ and $\mathbf{H}$ into (40), where care is taken to ensure positive definiteness of $[\mathbf{C}_0^Z - \mathbf{V} - \mathbf{R}]$ and $\mathbf{J}\hat{\mathbf{Q}}\mathbf{J}'$.

Estimation of State Matrix B

It can be shown (Wikle, 1996) that

$$\mathbf{B} = \mathbf{C}_1^Z \mathbf{J}' \left(\mathbf{J} \left[\mathbf{C}_0^Z - \mathbf{V} - \mathbf{R} \right] \mathbf{J}' \right)^{-1} \tag{42}$$

We can then obtain the estimate $\hat{\mathbf{B}}$ by substituting estimates of $\mathbf{C}_1^Z, \mathbf{C}_0^Z, \mathbf{V},$ and $\mathbf{R},$ into (41). Care must be taken to assure that $(\mathbf{J}[\mathbf{C}_0^Z - \mathbf{V} - \mathbf{R}]\mathbf{J}')$ is positive definite.

Estimation of $\mathbf{C}_0^Z$, $\mathbf{C}_1^Z$, and $\mathbf{C}_0^Y$

As shown above, in order to obtain estimates of the Kalman-filter model parameters, we need estimates of the covariance matrices $\mathbf{C}_0^Z, \mathbf{C}_1^Z,$ and $\mathbf{C}_0^Y.$ We estimate these matrices by MOM. In particular, let

$$\hat{c}_0 \left(Z(\mathbf{s}_i), Z(\mathbf{s}_j) \right) \equiv \frac{1}{T} \sum_{t=1}^{T} \left(Z(\mathbf{s}_i; t) - \hat{\mu}_Z \right) \left(Z(\mathbf{s}_j; t) - \hat{\mu}_Z \right) \tag{43}$$

where

$$\hat{\mu}_Z \equiv \frac{1}{nT} \sum_{i=1}^{n} \sum_{t=1}^{T} Z(\mathbf{s}_i; t) \tag{44}$$

is the estimate of the "grand" mean associated with the Z process. Then,

$$\hat{\mathbf{C}}_0^Z \equiv \left[\hat{c}_0 \left(Z(\mathbf{s}_i), Z(\mathbf{s}_j) \right) \right]_{i, j=1,\ldots,n} \tag{45}$$

Similarly, the (i,j)-th element of the lag-one covariance matrix is estimated by

$$\hat{c}_1 \left(Z(\mathbf{s}_i), Z(\mathbf{s}_j) \right) \equiv \frac{1}{T-1} \sum_{t=2}^{T} \left(Z(\mathbf{s}_i; t) - \hat{\mu}_Z \right) \left(Z(\mathbf{s}_j; t-1) - \hat{\mu}_Z \right) \tag{46}$$

and then

$$\hat{\mathbf{C}}_1^Z \equiv \left[\hat{c}_1 \left(Z(\mathbf{s}_i), Z(\mathbf{s}_j) \right) \right]_{i, j=1,\ldots,n} \tag{47}$$

Finally, given an estimate $\hat{\mathbf{R}}$ in (33) and $\hat{\mathbf{C}}_0^Z$, we obtain the covariance matrix estimate of the Y process:

$$\hat{\mathbf{C}}_0^Y \equiv \hat{\mathbf{C}}_0^Z - \hat{\mathbf{R}} \tag{48}$$

where we must assure that $\hat{\mathbf{C}}_0^Y$ is positive definite.

PRECIPITATION OVER THE SOUTH CHINA SEA

We have chosen to apply our spatially descriptive, temporally dynamic model (SDTD) to a precipitation data set covering the South China Sea (SCS) region of southwestern Asia. Precipitation was chosen because it is a variable for which reliable spatial predictions are typically difficult to obtain. This difficulty arises because precipitation variability encompasses such a wide range of spatial and temporal scales, and because the physical characterization of the processes leading to these scales of variability and their interactions are relatively poorly understood. Specifically, we selected the SCS region because it is representative of the interaction of these scales of variability, and because this region includes effects from both mid-latitude and tropical weather systems (e.g., Chen and Chen, 1995). In addition, the lack of precipitation data over the SCS provides a good test of the SDTD model's ability to predict in regions with sparse spatial sampling.

Data

Monthly precipitation data were obtained from the National Climatic Data Center (NCDC) operated by the National Oceanic and Atmospheric Administration (NOAA). The data were from the first version of the Global Historical Climatology Network (GHCN) surface baseline data set. This data set was created by a joint effort of the NCDC and the Carbon Dioxide Information Analysis Center at the Oak Ridge National Laboratory in 1992 for the purposes of providing a data set to be used to monitor and detect climate change (Vose et al., 1992). Over 7,500 precipitation stations are included in the GHCN, with data extending back into the 1600s. All of the data have a quality-control flag that gives an indication of possible serial and spatial continuity problems.

We selected 135 stations around the SCS for the period 1959 to 1988, as shown in Figure 17.1. With the exception of three stations, all of the stations chosen for this analysis had high quality data for at least 85% of the monthly time periods between January 1959 and December 1988. The three stations, all on the coast of Vietnam, did not meet this 85% criterion because the data were not available after 1975. Nevertheless, these stations were included because it was thought that their presence on the western coast of the SCS would help define the EOF basis functions used in the analysis.

The goal of this analysis is to predict the monthly precipitation over the SCS at locations and times where there are no data. We have selected a prediction grid that covers the majority of the SCS and that has uniform grid point separations of 2° in latitude and longitude. This prediction grid is shown in Figure 17.1.

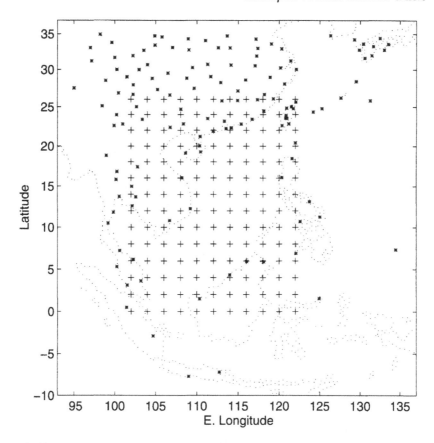

Figure 17.1. Observation locations (∗) and prediction grid (+) for the South China Sea precipitation example.

Exploratory Data Analysis

Monthly precipitation variability is often a function of the mean precipitation amount. To search for this mean-variance relationship, the following diagnostic methodology is adopted. A plot of the natural log of the mean versus the natural log of the variance of precipitation (time-averaged at each observation location) is considered (Figure 17.2). The linear relationship evident in this plot is quite striking given that the data are "real" and have not been generated synthetically. The slope and intercept from this empirical relationship leads to an approximate power of the mean model given by:

$$\hat{\sigma}^2(\mathbf{s}_i) \approx 3.00 \; \hat{\mu}(\mathbf{s}_i)^{1.55} \tag{49}$$

where $\hat{\sigma}^2(\mathbf{s}_i)$ and $\hat{\mu}(\mathbf{s}_i)$ are the estimated variance and mean at data location $\mathbf{s}_i$, respectively. The fact that the variance is a function of the mean invalidates the homogeneity of variance assumptions inherent in the SDTD model development. Therefore, we must perform a variance stabilizing transformation such that transformed data have constant variance (e.g., Snedecor and Cochran, 1989, pp. 286–287). This procedure is outlined below.

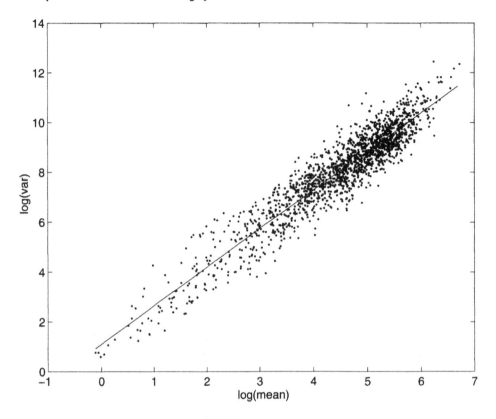

Figure 17.2. Natural log of precipitation (mm) mean vs. the natural log of the precipitation variance and the weighted least squares fit. Means and variances are calculated over time at each observation location.

Variance Stabilizing Transformation and Data Preparation

Assume we are given data X such that the variance is dependent on the mean in the following sense:

$$\sigma^2(\mu) = k\,\mu^\beta \tag{50}$$

where $\sigma^2(\mu)$ is the variance of X, μ is the mean of X, and k and β are unknown parameters. We then seek the transformation $f(X)$ that gives constant variance. It can easily be shown that expanding $f(X)$ in a Taylor series about μ and requiring $\mathrm{var}[f(X)]$ to be constant leads to the following relationship:

$$f(X) \propto X^{1-\beta/2} \tag{51}$$

In our case, we have estimated β to be 1.55, and we let k equal 1.0, so an appropriate variance stabilizing transformation is

$$f(X) = X^{0.23} \tag{52}$$

which is very close to the familiar 4-th root transformation.

In practice, after applying the transformation (52) to the precipitation data, we must remove the seasonal mean effect. This is done by calculating the time mean for each of the 12 months of the year, for each station. Then, the appropriate monthly mean is subtracted from the transformed data. These means must also be estimated at locations where we do not have data if we are to predict at such locations. We can obtain these estimates by applying the same pregridding spatial prediction machinery used for estimating the basis functions at locations where we do not have data.

After running the Kalman filter implementation of the SDTD model, we must transform the predictions and MSPEs back to the original scale. It is well known that such a transformation induces a bias in the prediction (e.g., Cressie, 1993, pp. 135–138). Therefore, we must correct for this bias. In our case, recall that we have defined the Kalman filter predictor in (21), (22) to be $\hat{Y}(\mathbf{s};t|t)$. Also, in the derivation of the SDTD model we assumed the data were given by observations on the Z process, which is taken to have zero mean and homogeneous variance. Now, we further assume that these data Z were obtained through a variance stabilizing transformation on the original data X [i.e., $Z = f(X)$] and that we are interested in predicting a smooth process W on the original scale, corresponding to smooth process Y on the transformed scale. That is, we would like to obtain $\hat{W}(\mathbf{s};t|t)$ from the Kalman filter prediction $\hat{Y}(\mathbf{s};t|t)$. Analogous to Cressie (1993, p. 137), we can use Taylor series expansions to obtain the following approximate relationship for the unbiased predictor:

$$\hat{W}(\mathbf{s};t|t) \approx \tilde{f}(\hat{Y}(\mathbf{s};t|t)) + (1/2)\tilde{f}''(\hat{\mu}_Y(\mathbf{s};t))\sigma_Y^2(\mathbf{s};t) \tag{53}$$

where $\tilde{f}(\ ) \equiv f^{-1}(\ )$, $\hat{\mu}_Y(\mathbf{s};t)$ is the estimate of the seasonal mean of the transformed data, and $\hat{\sigma}_Y^2(\mathbf{s};t)$ is the MSPE of the Y process as given by (24). It is also easy to show that the MSPE for the W process is given approximately by,

$$\hat{\sigma}_W^2(\mathbf{s};t) \approx \left\{\tilde{f}'(\hat{\mu}_Y(\mathbf{s};t))\right\}^2 \hat{\sigma}_Y^2(\mathbf{s};t) \tag{54}$$

In our case, the function $f(\cdot)$ is defined by (52), and so the function $\tilde{f}(\cdot)$ is defined as the inverse of (52), namely $\tilde{f}(Z) = Z^{1/0.23}$.

Implementing the SDTD Model

In order to implement the SDTD model, we must acquire basis functions for the observation network. First, we transform our map coordinates to an equal area projection. In particular, we use the cylindrical equal area projection described in Pearson (1990, pp. 129–132). This transformation is needed because the discretization of the continuous EOF analysis depends on weights that are based on the areas of the grid boxes in the network. Thus, we do not want a map projection with unequal areas, which would bias our determination of the basis functions. Then, basis functions are obtained by pregridding as described earlier.

We expect that the data are contaminated by measurement error since they are collected and disseminated by such a large variety of jurisdictions. To examine this, we plot the estimated empirical semivariogram in Figure 17.3. As is evident in this plot, there is a "nugget effect" associated with these data, thus implying microscale variability and/or measurement error. We assume that the entire nugget effect is due to measurement error, and choose a value of $\hat{\sigma}_\epsilon^2 = 0.03$.

Next, we examine the eigenvalues of the estimated matrix $\hat{C}_0^Y$ to help determine the truncation value K. We selected $K=50$ since the first 50 eigenvalues imply that the estimated matrix $\hat{C}_0^{Y_K}$ would then account for approximately 85% of the variance of $\hat{C}_0^Y$. Sensitivity analyses show that the results are relatively robust to the choice of this parameter, as long as it is reasonably large. As described earlier, estimates of the v-process covariance matrix $\hat{V}$ were found. Figure 17.4 shows the estimated v covariances at different lags, as well as the "bin" averages of those covariances. We note the large spike at lag zero, and relatively minor structure at higher lags. We did not feel there was sufficient structure in v to justify the extra computational effort required to deviate from the white-noise hypothesis given by (38). Then, we estimated $\hat{\sigma}_v^2 = 0.03$ (coincidentally the same as the measurement error estimate $\hat{\sigma}_\epsilon^2$) by the approach suggested earlier (see [39]).

SDTD Model Results with Precipitation Data

The SDTD model was run with the transformed precipitation data described above and the predictions were transformed back to the original scale for presentation. Figure 17.5 shows time series plots of the predictions at three locations for the 10 year period from January 1979 to December 1988: (a) a location on the southern coast of China (near 114°E, 22°N); (b) a location in the middle of the SCS (at 114°E, 12°N); and (c) a location along the northwestern coast of Borneo (near 114°E, 4°N). These locations were chosen because they provide a comparison of the predictions from the northern (data-rich) region, the central (no-data) region, and the southern (sparse data) region. In addition to the predicted precipitation (solid line), these figures show the monthly mean precipitation (dotted line), and Figures 17.5a and 17.5c show the observed precipitation data (dashed line). In each case, the predictions appear to have captured the appropriate structure in the time variability of precipitation. Although some of the extreme peaks are clearly missed by the predicted time series (e.g., the extreme event in 1982 in Figure 17.5a), in general, the predictions are able to capture the "direction" of the deviation from the seasonal mean. Although not shown, the Kalman filter also predicts successfully missing values in the time series.

Note that our interest in this example is spatiotemporal filtering of noisy observations and spatial prediction within the SCS region at all observation times. We are *not* interested in temporal prediction beyond the observation times. However, the Kalman filter formulation can provide such predictions via the one-step ahead prediction Equations 18–21 and similar estimates for the Y-process (e.g., Wikle and Cressie, 1997). Although such predictions might be useful in monthly weather prediction applications, they will not be discussed here.

We now consider the prediction over the grid covering the SCS (shown in Figure 17.1). For illustration, we present the results for 1979. This year was very active meteorologically in the region, with a strong SCS monsoon, Mei-Yu front, and 30–60 day oscillation (e.g., see Chen and Chen, 1995). Figure 17.6 shows surface plots of the pre-

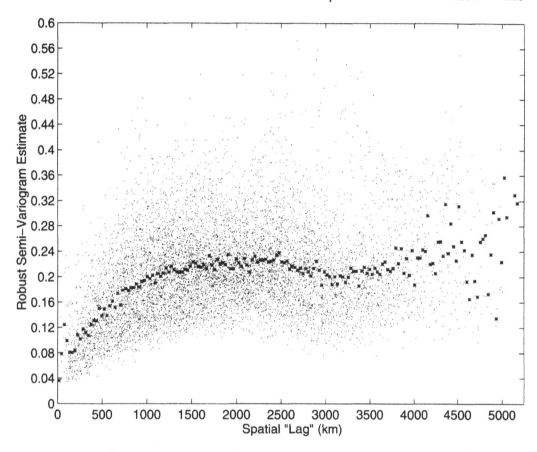

Figure 17.3. Robust semivariogram estimate of the transformed precipitation data (·) and "bin" averages (*).

dicted precipitation for April through September of 1979. The corresponding square root of the MSPE is shown in Figure 17.7. Note that, in April, the predicted precipitation shows two maxima, one in the north and one in the south, with a minimum over the SCS. The root MSPE plot corresponds similarly to these maxima and minima, as expected, since (54) shows that the transformation of the MSPE is dependent on the seasonal mean. Thus, areas with more mean precipitation will have larger MSPEs, and areas with less mean precipitation will have smaller MSPEs. The northern peak corresponds to the convection associated with the Mei-Yu front and the southern peak corresponds to that associated with the Intertropical Convergence Zone (ITCZ). Notice that throughout the spring (April, May) months, the central SCS shows a relative minimum in predicted precipitation. By July, the ITCZ has begun to move northward, leading to larger precipitation amounts over the central SCS. After reaching its maximum intensity in August, the ITCZ begins to retreat to the south. Although not shown here, by November the ITCZ is firmly entrenched in the southern region, and the north is very dry. This seasonal migration is also apparent in satellite-derived precipitation estimates for the same time period (e.g., Chen and Chen, 1995). Thus, the SDTD model appears to capture the dynamic evolution of the precipitation over the data-sparse SCS.

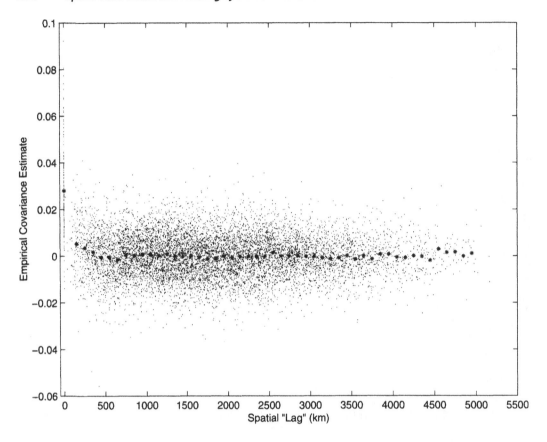

Figure 17.4. Estimated covariances of the ν process for the transformed precipitation process (·) and "bin" averages (*).

CONCLUSION

We have presented a spatiotemporal statistical model that considers the influence of both spatial and temporal variability. The model is temporally dynamic in that it exploits the unidirectional flow of time in an autoregressive framework, and is spatially descriptive in that no causative interpretation is associated with its spatially colored noise. The inclusion of a measurement equation naturally leads to the development of a spatiotemporal Kalman filter. The Kalman filter implementation allows us to predict in time and in space, and to account for missing data. The model was applied to 30 years of monthly precipitation data from the South China Sea region of Asia, and seems to capture the dynamic evolution of the spatial processes associated with the precipitation in this region.

There is much that can be done to further the development of this approach. This includes the investigation of alternative basis functions, extensions to multivariate space-time fields (i.e., temperature and precipitation, wind speed and direction, and so forth), and the inclusion of additional time lags in the model (3).

Finally, a promising direction, made possible by recent advances in Markov Chain Monte Carlo techniques, concerns the implementation of an SDTD model in a hierarchical Bayesian framework. Given that the Kalman filter can be interpreted from a Bayesian perspec-

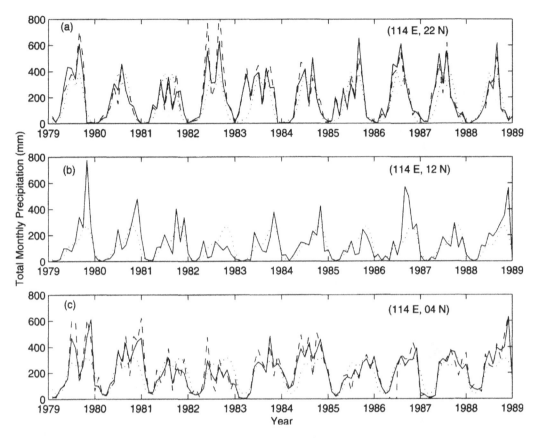

Figure 17.5. Time series plots for three locations over the 10-year period from January 1979 to December 1988: (a) a location on the data-rich southern coast of China (near 114°E, 22°N), (b) a location with no nearby data in the middle of the South China Sea (114°E, 12°N), and (c) a location in the sparse-data region along the northwestern coast of Borneo near (114°E, 4°N). The predicted precipitation (mm) is indicated by a solid line, the original observational data (mm) (for [a] and [c]) by a dashed line, and the estimated seasonal mean precipitation (mm) by the dotted line.

tive, the estimation of model parameters given the data, as applied in this presentation, corresponds to an empirical Bayes approach. As mentioned previously, when model parameters are estimated, the (estimated) Kalman filter no longer yields the optimal predictor (i.e., conditional expectation given by [6]). However, we could take the additional step and assign prior distributions to the model parameters. This is the hierarchical Bayesian approach and yields the optimal solution, subject to the assumptions placed on the prior distributions. Recently, Wikle et al., (1998) presented such a hierarchical Bayesian space-time model in an analysis of monthly temperature data and demonstrated it to be very promising.

ACKNOWLEDGMENTS

The authors wish to thank Professor Mike Chen for his helpful discussions regarding the South China Sea precipitation example. The first author was supported by the U.S. Depart-

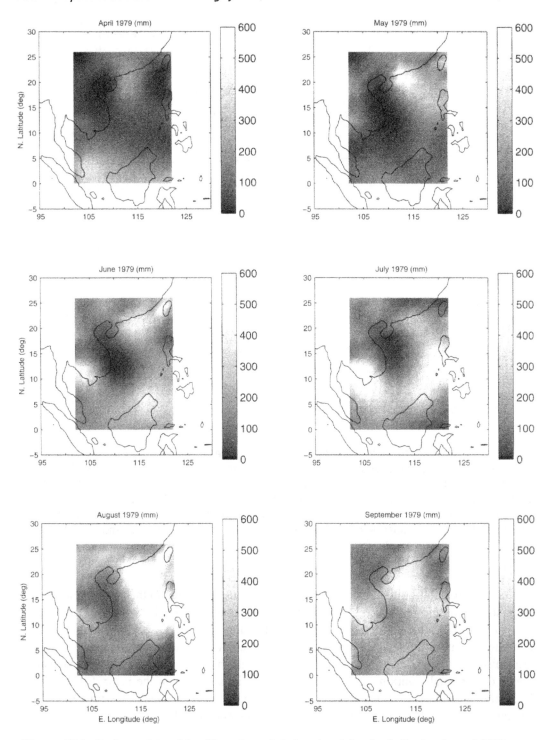

Figure 17.6. Surface plots of the filtered precipitation (mm) for April–September of 1979.

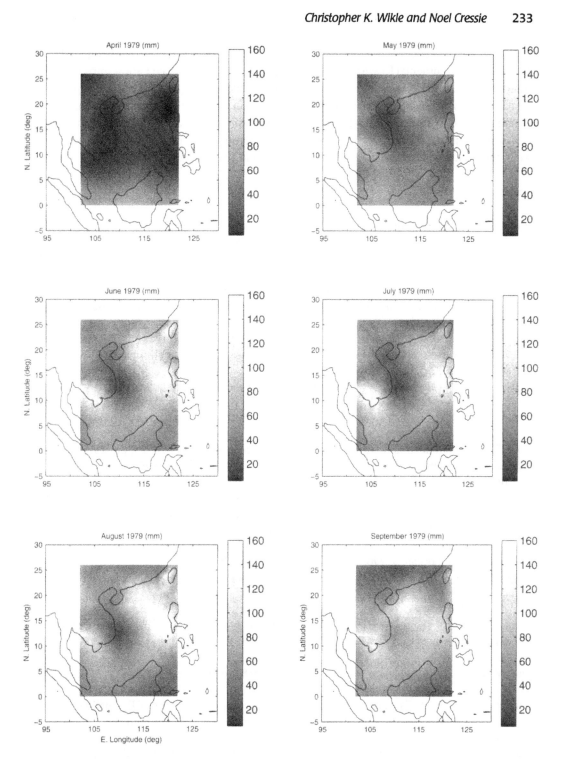

Figure 17.7. Surface plots of the filtered precipitation prediction error standard deviations (mm) for April–September of 1979.

ment of Energy, Office of Energy Research, Environmental Sciences Division, Office of Health and Environmental Research, under appointment to the Graduate Fellowships for Global Change administered by Oak Ridge Institute for Science and Education. Further support was provided by the NCAR Geophysical Statistics Project, sponsored by the National Science Foundation under Grant DMS93-12686. N. Cressie's research was partially supported by the U.S. Environmental Protection Agency under cooperative agreement CR 822919-01-0 between the U.S. Environmental Protection Agency and Iowa State University.

REFERENCES

Bilonick, R.A. Risk qualified maps of hydrogen ion concentration for the New York state area for 1966–1978, *Atmos. Environ.*, 17, pp. 2513–2524, 1983

Buell, C.E. Integral equation representation for factor analysis, *J. Atmos. Sci.*, 28, pp. 1502–1505, 1972.

Buell, C.E. The topography of empirical orthogonal functions, *Preprints, Fourth Conference on Probability and Statistics in the Atmospheric Sciences*, American Meteorological Society, 1975, pp. 188–193.

Cane, M.A., A. Kaplan, R.N. Miller, B. Tang, E.C. Hackert, and A.J. Busalacchi. Mapping tropical Pacific sea level: data assimilation via a reduced state space Kalman filter, *J. Geophys. Res.*, 101, pp. 22599–22617, 1996.

Chen, T.-C. and J.-M. Chen. An observational study of the South China Sea monsoon during the 1979 summer: Onset and life cycle, *Mon. Weather Rev.*, 123, pp. 2295–2318, 1995.

Cohen, A. and R.H. Jones. Regression on a random field, *J. Am. Stat. Assoc.*, 64, pp. 1172–1182, 1969.

Cohn, S.E. An introduction to estimation theory, *J. Meteorol. Soc. Japan*, 75, pp. 257–288, 1997.

Cressie, N. Comment on "An approach to statistical spatial-temporal modeling of meteorological fields" by M.S. Handcock and J.R. Wallis, *J. Am. Stat. Assoc.*, 89, pp. 379–382, 1994.

Cressie, N.A.C. *Statistics for Spatial Data, Revised Edition*, John Wiley & Sons, New York, 1993.

Creutin, J.D. and C. Obled. Objective analysis and mapping techniques for rainfall fields: An objective comparison, *Water Resour. Res.*, 18, pp. 413–431, 1982.

Davis, R.E. Predictability of sea surface temperature and sea level pressure anomalies over the North Pacific ocean, *J. Phys. Oceanogr.*, 6, pp. 249–266, 1976.

Eynon, B.P. and P. Switzer. The variability of rainfall acidity, *Can. J. Stat.*, 11, pp. 11–24, 1983.

Freiberger, W. and U. Grenander. On the formulation of statistical meteorology, *Rev. Int. Stat. Inst.*, 33, pp. 59–86, 1965.

Fukumori, I. and P. Malanotte-Rizzoli. An approximate Kalman filter for ocean data assimilation: An example with an idealized Gulf Stream model, *J. Geophys. Res.*, 100, pp. 6777–6793, 1995.

Ghil, M. Advances in sequential estimation for atmospheric and oceanic flows, *J. Meteorol. Soc. Japan*, 75, pp. 289–304, 1997.

Ghil, M., S.E. Cohn, J. Tavantzis, K. Bube, and E. Isaacson. Applications of Estimation Theory to Numerical Weather Prediction, *Dynamic Meteorology: Data Assimilation Methods*, L. Bengtsson, M. Ghil, and E. Källén, Eds., Springer-Verlag, New York, 1981, pp. 139–224.

Guttorp, P. and P.D. Sampson. Methods for Estimating Heterogeneous Spatial Covariance Functions with Environmental Applications, *Environmental Statistics*, Handbook of Statistics, Vol. 12, G.P. Patil and C.R. Rao, Eds., North-Holland, Amsterdam, New York, 1994, pp. 661–689.

Handcock, M.S. and J.R. Wallis. An approach to statistical spatial-temporal modeling of meteorological fields (with discussion), *J. Am. Stat. Assoc.*, 89, pp. 368–390, 1994.

Hastie, T.J. and R.J. Tibshirani. *Generalized Additive Models*, Chapman and Hall, London, 1990.

Host, G., H. Omre, and P. Switzer. Spatial interpolation errors for monitoring data, *J. Am. Stat. Assoc.*, 90, pp. 853–861, 1995.

Huang, H.C. and N. Cressie. Spatio-temporal prediction of snow water equivalent using the Kalman filter, *Computational Stat. Data Anal.*, 22, pp. 159–175, 1996.

Kalman, R.E. A new approach to linear filtering and prediction problems, *J. Basic Eng. (ASME)*, 82D, pp. 35–45, 1960.

Karl, T.R., A.J. Koscielny, and H.F. Diaz. Potential errors in the application of principal component (eigenvector) analysis to geophysical data. *J. Appl. Meteorol.*, 21, pp. 1183–1186, 1982.

Le, D.N. and A.J. Petkau. The variability of rainfall acidity revisited. *Can. J. Stat.*, 16, pp. 15–38, 1988.

MathWorks. *MATLAB New Features Guide (December 1996)*, MathWorks, Inc., 1996.

Obled, C. and J.D. Creutin. Some developments in the use of empirical orthogonal functions for mapping meteorological fields, *J. Climate Appl. Meteorol.*, 25, pp. 1189–1204, 1986.

Papoulis, A. *Probability, Random Variables, and Stochastic Processes*, McGraw-Hill, Inc., New York, 1965, pp. 457–461.

Pearson, F. *Map Projections: Theory and Applications*, CRC Press, Boca Raton, FL, 1990.

Preisendorfer, R.W. *Principal Component Analysis in Meteorology and Oceanography*, Elsevier, Amsterdam, 1988.

Sampson, P.D. and P. Guttorp. Nonparametric estimation of nonstationary spatial covariance structure, *J. Am. Stat. Assoc.*, 87, pp. 108–119, 1992.

Sandwell, D.T. Biharmonic spline interpolation of GEOS-3 and SEASAT altimeter data, *Geophys. Res. Lett.*, 14, pp. 139–142, 1987.

Shriver, J.F. and J.J. O'Brien. Low-frequency variability of the equatorial Pacific ocean using a new pseudostress data set: 1930–1989, *J. Climate*, 8, pp. 2762–2786, 1995.

Smith, T.M., R.W. Reynolds, R.E. Livezey, and D.C. Stokes. Reconstruction of historical sea surface temperatures using empirical orthogonal functions, *J. Climate*, 9, pp. 1403–1420, 1996.

Snedecor, G.W. and W.G. Cochran. *Statistical Methods*, Eighth ed., Iowa State University Press, Ames, IA, 1989, p. 503.

Stein, M. A simple model for spatial-temporal processes, *Water Resour. Res.*, 22, pp. 2107–2110, 1986.

von Storch, H. and F.W. Zwiers. *Statistical Analysis in Climate Research*, Cambridge University Press, 1999.

Vose, R.S., R.L. Schmayer, P.M. Steurer, T.C. Peterson, R. Heim, T.R. Karl, and J.K. Eischeid. The Global Historical Climatology Network: Long-term monthly temperature, precipitation, sea level pressure, and station pressure data, NDP-041, Carbon Dioxide Information Analysis Center, Oak Ridge National Laboratory, Oak Ridge, TN, 1992.

Wikle, C.K. Spatio-Temporal Statistical Models with Applications to Atmospheric Processes, Dissertation presented to Iowa State University in partial fulfillment of the requirements for the degree of Doctor of Philosophy, 1996.

Wikle, C.K., L.M. Berliner, and N. Cressie. Hierarchical Bayesian Space-Time Models, *Environmental and Ecological Statistics*, 5, pp. 117–154, 1998.

Wikle, C.K. and N. Cressie. A Dimension Reduction Approach to Space-Time Kalman Filtering, Preprint Number 97-24, Statistical Laboratory, Iowa State University, Ames, IA, 1997.

Index

Printed and bound by CPI Group (UK) Ltd, Croydon, CR0 4YY

24/10/2024

01778287-0018